Infusing Technology into the Classroom:

Continuous Practice Improvement

Catherine C. Schifter
Temple University, USA

Information Science Publishing
Hershey • New York

Acquisition Editor:	Kristin Klinger
Senior Managing Editor:	Jennifer Neidig
Managing Editor:	Jamie Snavely
Assistant Managing Editor:	Carole Coulson
Development Editor:	Kristin M. Roth
Editorial Assistants:	Ross Miller and Deborah Yahnke
Copy Editor:	Jennifer Young
Typesetter:	Carole Coulson
Cover Design:	Lisa Tosheff
Printed at:	Yurchak Printing Inc.

Published in the United States of America by
Information Science Publishing (an imprint of IGI Global)
701 E. Chocolate Avenue
Hershey PA 17033
Tel: 717-533-8845
Fax: 717-533-8661
E-mail: cust@igi-global.com
Web site: http://www.igi-global.com

and in the United Kingdom by
Information Science Publishing (an imprint of IGI Global)
3 Henrietta Street
Covent Garden
London WC2E 8LU
Tel: 44 20 7240 0856
Fax: 44 20 7379 3313
Web site: http://www.eurospanbookstore.com

Library of Congress Cataloging-in-Publication Data

Schifter, Catherine.
Infusing technology into the classroom : continuous practice improvement / by Catherine C. Schifter.
p. cm.
Includes bibliographical references and index.
ISBN-13: 978-1-59904-765-2 (hardcover)
ISBN-13: 978-1-59904-767-6 (e-book)
1. Educational technology. 2. Education--Effect of technological innovations on. 3. Teaching. I. Title.
LB1028.3.S345 2008
371.33--dc22

2007038034

British Cataloguing in Publication Data
A Cataloguing in Publication record for this book is available from the British Library.

To Stephan, Diane, Chester, and Sasha

Infusing Technology into the Classroom:

Continuous Practice Improvement

Table of Contents

Foreword

In the 1840s, school reformers fretted that teachers were making scant use of an incredible new instructional technology. The chalkboard had been introduced into the nation's emerging common schools a decade or two earlier, but classroom visits revealed little if any use. It seemed that the students knew how to use the device better than the teachers did. To prepare teachers to take advantage of this innovative technology, normal schools (early teacher training colleges) introduced courses on chalkboard use. Textbooks to support those courses were written, early forms of professional development were instituted, and how-to guides created. Why were chalkboards, eventually such a natural part of the classroom routine, so difficult to embrace?

When we think about the elements that make today's instructional technology devices—computers, interactive whiteboards, handhelds, and so on—complex, we tend to dwell on technical issues. How do I connect to the Internet? Will I break it if I press this button? What does Exit mean? How do I turn it off? How do I turn it on? Early chalkboard users may have confronted questions about best use. Is it better to use block letters or cursive? How much detail should I write? Should I put all my notes on the board or just the most important ones? What happens when I turn my back on the students? But the device itself couldn't have been that complicated to use technically. Touch a piece of chalk to the board, move it around, and you're on your way. Simple, right? So why didn't teachers use them?

The reformers who introduced chalkboards to the country's emerging common schools drew their inspiration from what they had witnessed in university teaching. There they saw college professors dynamically illustrating their lectures on large pieces of slate, bringing their words to life with timely illustrations and notes matched perfectly to their unfolding lessons. Such instructional dynamism clearly also should be in the hands of teachers working with younger children. However, the one-room schoolhouses that dominated early 19th century American common education weren't structured to match that pedagogical vision. A one-room schoolhouse of the time could easily require a teacher to manage groups of children, sometimes numbering 100 or more, of all ages and abilities, covering all subjects, at once. Rarely, under those circumstances, did a teacher teach everyone at the same time. Instead, teachers relied on recitations and a monitorial system that might today be called peer-assisted drill and practice.

At the time, the chalkboard offered a much more active and engaging classroom experience than the dominant norm. But the way teachers taught and the context in which that teaching happened both had to change for the chalkboard promise to be realized. The technical part of learning to use the chalkboard was easy; the challenge came with the pedagogical demands accompanying it.

The shift from lumping all children together in a one room schoolhouse to separating those children into grades by age had a huge impact on the form of classroom teaching. With the change in structure, the vision of dynamic, teacher-led (and chalkboard-fed) instruction could now match a school infrastructure that placed kids of similar ages and abilities together. The chalkboard, once thought to be too complicated for teachers, quickly became an essential tool of the trade, and one, some might argue, that rarely reached its inspirational potential over the last 150 years. In fact, today's critics of the educational norm might see the the entrenched chalkboard as emblematic of a dominant chalk-and-talk pedagogy that neglects student interests and capabilities and blocks deep, self-constructed understanding.

Hopefully this history lesson on the chalkboard can offer some insight into current technological innovations in K-12 schooling. The digital devices of the past few decades have certainly posed technical hurdles for teachers, but, like chalkboards, the bigger obstacles to acceptance may well be pedagogical. The promise of 21st century educational technologies once again threatens the instructional norm. What does it mean to reach toward the promise associated with computers and the Internet within the current structure of K-12 schooling?

In this book, Catherine Schifter captures the stories within a system that grappled with that question. As you read about the people and processes, I urge you to look beyond the technology itself. The tales in the pages ahead are not about databases, simulations, webquests, and educational productivity tools. They are about change. And lessons learned about change live long after the technology has faded.

David Dockterman

*Dr. **David Dockterman** is vice president and chief academic officer at Tom Snyder Productions, where he has been developing award-winning, research-based educational software for the last 25 years. Among his team's scores of programs are Decisions, Decisions; Thinking Reader; FASTT Math; and TimeLiner. David authored the books Great Teaching in the One Computer Classroom and Weaving Technology into Your Teaching. He also co-created and co-wrote Science Court, the highly acclaimed animated TV show that ran for three years on ABC's Saturday Morning. A former classroom teacher, David is an adjunct lecturer on education at the Harvard Graduate School of Education, where he earned his Ed.D. He served nine years as an elected member of his local school committee and has dedicated himself to supporting classroom teaching and the successful integration of technology into schools.*

Preface

Schools have been part of the American scene since the beginning of the American scene. They did not have the look and feel of a 21st century school, often geared more toward religious instruction than the study of literature and mathematics. Often in the 19th century Western towns, the school was a room in someone's house or a one-room school house. Perhaps this also was the case more often than not, but that is not what this book is about. The point in mentioning that schools are not new to America is that schooling has a long tradition, not necessarily for all who lived in America, but a long tradition nonetheless.

Teaching also has a long tradition; it always has been a time honored profession. While the mission may have grown into enculturation of immigrants into the "American way," whatever that means, students have learned to read, write, do arithmetic, and think about important ideas on the way to becoming adults and productive members of society.

Teaching, for many decades, consisted of the teacher telling the students what they needed to know and students "learning" what had been established by someone as the curriculum, or what they "needed to know." Students were empty vessels waiting to be filled with facts and information that they were then to turn into knowledge by way of assignments and examinations. Whether that transformation occurred was a mystery to many, but the "plan" was for it to happen.

In addition, technology is not new to schools, but has had a history of both success and failures. Cuban (1986) documented clearly that each time technology or machines were introduced into classrooms, with the notion that they were going to revolutionize education, nothing changed. The fact was that the film projectors had to be scheduled in advance, not used serendipitously, thus making them appear to be unavailable and difficult to use. And how many of the readers who lived through schooling in the 1950s remember the film breaking at the splice in the middle of the film, and thus ending that part of the lesson? There were problems for sure. Teachers were not technicians, and not all schools had an educational technologist who could fix what ever problems arose, whether the bulb burned out in the overhead projector or the film broke. Thus, teachers began to think of technology as more of a nuisance than a boon.

Then computers came into classrooms. Here again Cuban (1986, 2001) and Schofield (1995) chronicled that under use, or lack of use, of computers in classrooms was predictable and due to inertia and resistance by teachers. This was true of some teachers, especially in the late 1980s and early 1990s when computers were still unique in classrooms and not prevalent in homes, especially not in teachers' homes. But it is not fair to blame the teachers for lack of use of computers. For many reasons, the computer is one pedagogical option for teachers, *not* the only option.

Teachers are not known as risk takers. Teachers often teach the way they were taught, which is apt to be the way they like to learn. They are in charge in their classrooms, not the students. That is a very important fact to remember. Changing the dynamics or social structure in the classroom such that the teacher may no longer be the most important person in the room is tantamount to heresy! And yet for computers, in particular, to be effective in supporting student learning in classrooms, where the ratio of students to computers is likely to be 30 to 1 at worst and 4 to 1 at best, the teacher must be comfortable with ceding some control to students, which they may think will lead to chaos. Allowing different activities to happen at the same time with not all students listening to the teacher is not the paradigm with which most teachers grew up, nor do most administrators understand. I will always remember a colleague relating a story about his department chair coming to review his teaching in one of his classes. The students were in groups working on different projects and my colleague was moving between the groups assisting and keeping them on track. The department chair, uncomfortable with the apparent noise and activity, told my colleague that he would come back another time when he was really "teaching.." And, he was not using computers, but active learning in cooperative working groups! So the vision of what is appropriate in classrooms makes a big difference on what happens there.

Schofield (1995) describes a school from the mid-1980s and how technology was used little if at all. She notes that there was training available for teachers at a "Teachers' Center" (p. 117) on many topics, including computers. Unfortunately, this training was away from the classroom and divorced from the reality of the teacher's students and teaching needs. Such training, unless reinforced back in the school, routinely fails to make a difference in classroom teaching practices. In addition, such training is only about the hardware and possibly specific software that may or may not be available in the classroom, thus transfer of skills may be difficult at best. And for those teachers who are concerned about maintaining the aura of authority in the classroom, showing a lack of knowledge or lack of expertise would be difficult to handle. One thing that is not taught in colleges of education is how to say, "I don't know, but let's find out together."

Computers have continued to develop and become more powerful tools than was ever dreamed of when they came into schools in the late 1970s. And because they continue to change, the problem of keeping up with the new software, operating systems, and tools can be awkward and intimidating. It is much easier to say, as one of the teachers quoted in Schofield's 1995 book, *Computers and Classroom Culture*, put it, "It [the computer] didn't do anything I couldn't do easier and cheaper on the blackboard." (p. 103) This is the ostrich approach to change—stick with what you know, stick your head in the sand, and perhaps _____ (insert the innovation du jour) will just go away and I can keep doing what I've been doing.

Well, it is true that technology, including computers, often arrives in classrooms not at the request of the teacher. It continues to be true that most of the formal and professional development training teachers receive about computers and teaching occurs outside the classroom. And it is true that there are many computers in many classrooms that are not used, or considered to be underused by critics of technology use. These facts do not negate the considerable impact computers and other technologies have on schools in spite of barriers and inertia by many.

About this Book

This book is about one staff development program designed to help teachers change their teaching practices through a professional development program developed initially in the Philadelphia school district in conjunction with the IBM® Corporation, called Continuous Practice Improvement (CPI). The purpose of this book is for readers to identify with one or more of the teachers whose stories are chronicled here and to understand that using computers in teaching and learning is not about technology, but about good teaching and classroom management. These were brave teachers at various stages of their careers who wanted to be 21st century teachers, recognizing that computers were not going away, and they needed to help their students be 21st century citizens.

What made this program unique was not that teachers were given technology for their classrooms, but that the program adopted the medical training model of "resident" teachers, like resident doctors, who worked with other teachers in their own environment to "show" these teachers how they practice teaching that includes technology/computers as just part of the process. These resident teachers provided guided experiences for the novice-technology-using teacher in practical ways that allowed the novice-technology-using teacher to envision how this would work in their own classroom. Then they went back to their classroom and made the changes, tried out the new methods, revised the processes in collaboration with the Resident Teacher and other novice-technology-using teachers participating in the process in other classrooms, thereby changing their approach to teaching.

The Continuous Practice Improvement program, or CPI, was not about learning how to use a computer or specific software. Yes, the novice-technology-using teacher needed to become comfortable with computing and computers, but that was not the main point. It was not about using computers to teach reading or mathematics or social studies or science or anything in any particular classroom. CPI was about good teaching in general, classroom management practices that allow for different activities occurring in a classroom at the same time without chaos taking the reins. Computers were just part of the process.

One example of the success of the CPI program is a fourth grade teacher, who you will meet in Chapter IX, from an inner city school that is located in a disadvantaged area of the city. When she came the first day of training she was absolutely convinced that she could not use a computer and in fact she was going to break it. She came with others from her school who had encouraged her to participate so she could get a new computer for her classroom, but that was the only reason she was there. When she received her loaner-laptop, she could not get it to turn on, therefore she had broken it. Then she could not open the Web browsing software, so she had broken it. One of her colleagues tried to help her, but she was convinced she was not going to be successful. The next session she was a bit more comfortable with the computer and was beginning to learn the software. When I visited her classroom six months later, she had two computers in her classroom in a center for her students to use for research projects. The students were using specific Web portal options that she set up for them to do research on different states. Each child was assigned a state on which to research and write a short paper. They, the students, then were going to create a joint project so they would learn from each other, not from the teacher. She was comfortable being the guide, the overseer of the projects, but not needing to be the direct deliverer of information. Not only was she now sure she could not break the computer, but she was sure her students were learning valuable lessons because they had access to the computer and information found on the Internet. Could they have done the projects without the computer? In this case no, because the classroom and school did not have enough resources. So the computer brought something to her classroom that she could not have achieved before, and would never have imagined doing before the CPI experiences. And the students were excited about what they were doing, learning, and creating. Is that not what teaching should accomplish?

But that is just one example. This book is about the stories of these teachers and how this program changed their practices. Not all the stories are equal in length or weight. Some teachers changed more than others, but that is not the essential message. In this book, as previously noted, the author hopes that one of the stories will resonate with a reader who, like the fourth grade teacher above, believes computers are just too complicated and not necessary. They can not use computers in their classroom because it would be too disruptive. But then they will read about another teacher who teaches the same grade or perhaps one above or below the grade they teach, and read how that teacher changed their practices through CPI, and they will see that it might work for them. The problem will be that they do not have a CPI program in their school district. But nothing will be stopping them from working with others to set up such a program. It can be done.

How does this book contribute to knowledge about teachers and professional development? The Continuous Practice Improvement program was studied by myself and my research assistants for three years and was also studied by the researchers from the Center for Children and Technology of the Educational Development Center. While it has a successful track record overall, this history and process of the program has remained hidden from the public at large. The program that initially supported CPI was the Reinventing Education (RE) program of IBM, which was in its 11th year of existence in 2006 and had grown to a $75 million investment by IBM in 21 U.S. cities and 10 countries outside the U.S. Unfortunately, CPI was expensive and thus discontinued in Philadelphia, even though it was a successful program.

One contribution this book makes is in creating an amalgamated professional development model or lens for how districts and schools can determine and assess whether technology oriented professional development efforts can and will establish sustained change in classroom practices. The combination of a strong and useful training program with time to practice, effective and ongoing post-training technical and social support, and changes in classroom structures and values leads to sustained classroom changes in practice. This model uses all the lenses used in analysis of the teachers' stories to present a composite lens that district leaders, principals and teachers can use to guide professional development efforts.

Setting the Stage: The Philadelphia School District

In 2004, the School District of Philadelphia was the seventh largest in the nation based on enrollment, serving a racially and ethnically diverse student population. As of April 2004, the enrollment totaled 196,309 students, including early childhood programs like Head Start and preschool. Over 65 percent of the student population was classified as African-American, with over 14 percent Hispanic, over five percent Asian, over 14 percent White, and a small percent Native American.

There were 273 schools as of April 2004, including 175 elementary schools, which range in grades from Kindergarten through fourth grade to Kindergarten through eighth grade (and includes one school that was a Kindergarten through 12th grade school), and 43 middle schools in 2004. The district began to phase out most middle schools starting in fall 2005, meaning the configuration of many elementary schools changed to include middle grades or middle schools became fully functioning elementary schools. Thus, the number of elementary and middle schools as recorded in April 2004 was not the case in fall 2006, and this data was not available since the landscape was still changing. There were 43 neighborhood and magnet high schools, and five vocational-technical schools. As with changes in elementary and middle schools, the district administration was moving to reduce the size of many high schools by dividing them into smaller high schools and creating specialty high schools, such as the new high tech high school created in partnership with Microsoft Corporation that opened in fall 2006.

The district was organized into nine regions in April 2004, with the Center City region added in fall 2005. The Center City of Philadelphia experienced a surge in population over five years prompting the district to recognize the importance of highlighting and supporting the schools that served the Center City area, an attempt to retain the children of the Center City residents in public schools.

In order to maintain the dignity and anonymity of the teachers who volunteered to be interviewed for this book, their names have been changed to pseudonyms to mask their identities. In the end, after five years of offering the program under the direction of Ms. Maguire, 294 teachers (29 males, or 9 percent, and 265 females, or 91 percent) from 63 schools completed the program. The names of the schools where they taught are not given, but identified by the area of the city in which they exist along with some basic demographic information. The number of students in the schools will be presented as either "low" level of enrollments (between 300 and 500 students), "middle" level of enrollments (between 600 and 800 students), or "high" level of enrollments (between 900 and 1200 students), again to mask individual schools that stand out because of enrollments. In addition, the level of student attendance and mobility will be presented. District-wide attendance by students was expected to be at the 93 percent level, which is reached by most schools; however, we acknowledge that many children have competing pressures causing them to miss school for reasons other than illness. The same can be said about mobility of students across the district. One teacher described the student population in Philadelphia as "migratory," in that there are some students who move frequently from school to school. The reasons for this migratory behavior is vast and speculative, and not the point of this book; however, this statistic will be presented to help readers identify with the schools.

The Research Process for This Book

Working with the CPI program Director, teachers who participated in the CPI program in either the original design or revised design for implementation across the district and who were still teaching in the same schools as when they participated in the program were identified. They were contacted initially by e-mail or in person to gain permission to interview them for the book. Once permission was obtained, they were interviewed either over the phone or in their classrooms at a time convenient for them. The four basic questions asked were the following:

1. In what context were you teaching when you participated in the CPI program?
2. What were your personal conditions (feelings, hopes, needs, moral dispositions) that brought you to want to infuse technology into teaching?
3. What factors outside yourself (policies, school, teaching, other teachers) influenced your decision?
4. Considering over time, what has changed in your teaching practices since CPI?

The researcher took copious notes about the conversations and transcribed them within 24 hours of the interview. The transcript was e-mailed to the teacher for clarification and verification that the information was presented accurately and represented their thoughts properly. Teachers were given the opportunity to approve the transcription or edit it for clarity. Interview transcripts were grouped by grade level and theories of change in schools (see Chapter II) were used to analyze how the CPI program changed their practices. The following presents to the reader the contents of this book.

In the end, over 70 classroom teachers who participated in the CPI program were interviewed. This book presents 57 of these teachers who reported experiences that were different from each other. Of these 57 teachers, five were male (9 percent) and 52 were female (91 percent), which is equivalent to the population that participated in the CPI program overall. While there were attempts to interview many more CPI teachers, the ones presented in this book were randomly selected based on their availability and willingness to be interviewed, which may present a slight bias toward those teachers who were willing to talk about their CPI experiences. However, as the reader will find, these teachers were from different parts of Philadelphia, from different types of schools, but were willing to embrace change in their classrooms for the 21st century—the first step toward changing practice.

Teachers and Technology: Continuous Practice Improvement of Classroom Teaching and Learning

Section I: Promises, Promises: Microcomputers in Classrooms

Chapter I tells a very brief history of computing starting with Charles Babbage's Difference Machine. This is a brief history because the long history is available from many sources. Computers came into being to solve computing problems. With each generation of computers, the computers have gotten smaller, faster, and more efficient. The arrival of the Internet and subsequent World Wide Web revolutionized how we communicate around the world and have access to information. Overall, computers have changed the way people think about the size of the world, as exemplified by Thomas Friedman's book, *The World is Flat* (2005).

This chapter continues with a brief history of computing in schools, including the initial movement with computer-assisted instruction using mainframe options through universities. Again, this is a brief history because the complete history is available through many resources. Computers were initially used to mimic programmed instruction and task analysis from the 1950s and arose from the work of B.F. Skinner, among others. There were many different instructional systems created to support teaching, not the least of which was LOGO. LOGO appears to have had a major influence on some teachers' acceptance of computers into schools. How computers have impacted classrooms is discussed, which has not always been positive, but the research methods used have been inconsistent as well. Overall, teaching with computers has changed from teaching to program the computer to how the computer can facilitate learning in general.

Lastly, Chapter I presents a brief history of computing in Philadelphia schools. The Philadelphia story seems to have never been told. But this history sets the background for the CPI program. While the overall story of how computers infiltrated schools in Philadelphia is not unlike that of other large, urban cities, there were interesting components to this process. LOGO was a major force for those early adopters of computers in their classrooms. Now every school is wired for fast access to the Internet. In 2003, the district instituted an Internet-based instructional management system to support teaching and learning, giving each teacher from Kindergarten through eighth grade a laptop. Smartboard technology is coming into schools. Changes continue to occur with regard to computers in Philadelphia schools.

Chapter II presents theories of change in classroom practice, the ever present challenge to teachers and schools. Six approaches for evaluating how change occurs are presented, including Rogers theory of diffusion of innovations (1962, 1972, 1983, 1995, 2003), work by Fullan and Pomfret (1977) on determinants of implementation of innovations in schools, work by Berman and McLaughlin (1978) as part of the Rand Corporation study of federal change agent grants, the Concerns Based Adoption Model (CBAM) by Hord, Rutherford, Huling-Austin, and Hall (1987), and Fullan's three interactive factors that affect implementation of educational change (2001). These frames of reference will be used to assess the level of change in practice found in studying the teachers interviewed for this book.

Chapter III presents a brief history of professional development for teachers over the last thirty years, then moves to professional development for teachers to use computers. Given that professional development is all about effecting change in teacher practices, the discussion of how change happens from Chapter II is important to this discussion. This chapter is a brief history of professional development of teachers because there are legions of books, magazines, and journals dedicated to this topic. Here will be presented some of the issues that have been discussed over time, including whether or not professional development of teachers makes a difference.

Chapter IV relates the development of the CPI program through the IBM Reinventing Education project. Specifically, the chapter discusses how the teachers from the initial three schools worked with staff from IBM to develop and plan the CPI model of professional development, and then how this model was expanded district-wide. The program itself is described, including what was taught in the three Saturday training sessions, as well as what was intended to happen in the visitations with the resident teachers. One component that is important to the chapter is the director of the CPI program, Ms. Maguire. She was dedicated to the model, and still is. She was the driving force that made this model work for these teachers. This model was not about computing, *per se*, but about good teaching and classroom management. The key is having someone who champions the model and makes it work.

Section II: Teachers Voices: Early Childhood Teachers (K-2 Grades)

Chapter V presents seven Kindergarten teachers interviewed for this book. While not all the Kindergarten teachers who participate in CPI were interviewed, I firmly believe these are representative of how others' classroom practices changed. Chapter VI presents six first grade teachers interviewed for this book, and Chapter VII presents six second grade teachers. For each grade level, the impact of change theory and professional development will be discussed.

Section III: Teachers Voices: Elementary Grade Teachers (3-5 Grades)

Chapter VIII presents six third grade teachers interviewed for this book, while Chapter IX presents six fourth grade teachers and Chapter X presents seven fifth grade teachers interviewed for this book. As noted with Section II, this is a representative sample of the third, fourth, and fifth grade teachers who participated in the CPI professional development program. The impact of change theory and professional development on their practices will be discussed.

Section IV: Teachers Voices: Middle Grade Teachers (6-8 Grades)

Chapter XI presents six sixth grade teachers, while Chapter XII presents six seventh grade teachers and Chapter XIII presents seven eighth grade teachers. It is of interest to note that as the teachers were in higher grades, less change in classroom practices occurred and there were three instances of teachers with either no overt change in classroom practice or change in only personal use of computers.

Section V: Conclusion and Professional Development Model

Lastly, **Chapter XIV** presents an analysis of the level of change that appeared to have occurred in these teachers classrooms, as self disclosed. Change does not come easily or in a vacuum. In addition to the six lenses related to change presented in Chapter II, the Apple Classrooms of Tomorrow stages of technology adoption are used to analyze how these teachers grew from non-users of computers toward infusing computing into their classroom practices for teaching and learning. Thus, seven different, but related, lenses are used to analyze the level of change that occurred for these teachers. From this research and analysis, a theoretical lens is presented to inform teacher professional development to effect meaningful change in classroom practice as related to changes in uses of computers in classrooms.

References

Berman, P., & McLaughlin, M.W. (1978). *Federal programs supporting educational change, Volume VIII: Implementing and sustaining innovations*. Santa Monica, CA: Rand.

Cuban, L. (1986). *Teachers and machines: The classroom use of technology since 1920*. New York: Teachers College Press.

Cuban, L. (2001). *Oversold and underused: Computers in the classroom*. Cambridge, MA: Harvard University Press.

Friedman, T. L. (2005). *The world is flat: A brief history of the twenty-first century*. New York: Farrar, Straus, and Giroux.

Fullan, M. (2001). *The new meaning of educational change*. Third Edition. New York: Teachers College Press.

Fullan, M., & Pomfret, A. (1977). Research on curriculum and instruction implementation. *Review of Educational Research, 47*(1), 335-397.

Hord, S.M., Rutherford, W.L., Huling-Austin, L., & Hall, G.E. (1987). *Taking charge of change*. Alexandria, VA: Association for Supervision and Curriculum Development.

Rogers, E.M. (1962, 1972, 1983, 1995, 2003). *Diffusion of innovations,* Fourth edition. New York: The Free Press.

Schofield, J.W. (1995). *Computers and classroom culture*. New York: Cambridge University Press.

Acknowledgment

This book simply would not have been possible without Diane Maguire; her dedication to CPI and her assistance to me were invaluable. To all the Philadelphia School District teachers who participated in CPI, and especially those who gave of their time to be interviewed: a huge thank you. Thanks also to the past and present members of the Educational Technology Group of the District who so successfully implemented CPI and graciously gave their time to me.

Those who read the entire manuscript and provided their critique, comments, and suggestions deserve a special thank you. They were Patti Bateman, Aurora County Colorado Librarian and my sister; Steven E. Maffeo, Associate Director, McDermott Library, United States Air Force Academy; Bernice Harris. PhD, Vice President of Instruction, Community College of Aurora; Diane Melley, Director, IBM Corporate Community Relations; and lastly Stephan, my husband.

To my colleagues at Temple University—Concetta Stewart, Nina Gavula, Marc Lamont Hill, Diane Jass Kettlehut, Sarah-Kate Lavan, Bernie McGee, Aneta Pavlenko, and Tom Walker—my undying gratitude for their support, assistance, and encouragement.

A very special thank you to David Dockterman, Ed.D., Vice President and Chief Academic Officer, Tom Snyder Productions® for graciously penning the forward.

And lastly, to my publisher, IGI Global, and to the Editorial Assistants Deborah Yahnke and Ross Miller, a special thank you for all your assistance and for having the confidence in me to make this book a reality.

Catherine C. Schifter, PhD

Philadelphia, PA

August 28, 2007

Section I

Promises, Promises:
Microcomputers in Classrooms

There was a time when electricity was a novelty; something to be harnessed, to be understood. With time, not only did mankind harness electricity and understand it, but electricity became a mainstay of modern society. Today we cannot imagine our lives without light bulbs, refrigerators, and all the wonderful appliances that run on electricity. In the late 20th century the microcomputer came into being and was a toy for the wealthy, or mostly used in businesses. Those who imagined the power of computers in schools were few. Over time, as with electricity, computers have become part of the everyday landscape. But while computers have been in schools since the 1970s, teachers are still trying to figure out what to do with those "contraptions." Change is difficult, and change in teaching practices and norms seems to be even more difficult.

Today's modern computer has a long and distinguished history, but not so long for microcomputers in classrooms. Chapter I will first provide a brief history of computers in general, followed by a discussion of microcomputers in schools. Lastly, the history and impact of microcomputers in Philadelphia from the early 1980s through the early years of the 21st century will be presented to establish the background for the Continuous Practice Improvement program. Chapter II presents the theoretical lenses used throughout this book to help understand the impact of the Continuous Practice Improvement Program on the teachers presented in Section II, III, and IV. Chapter III reviews professional development in general, as well as professional development for integrating computers into classrooms. Chapter IV presents the development and implementation of the Continuous Practice Improvement (CPI) program in the School District of Philadelphia, setting the stage for the next three parts of the book which present teachers telling about their experiences with the CPI program and how it impacted their classroom practices immediately and over time.

Chapter I

A Brief History of Computers, Computing in Education, and Computing in Philadelphia Schools

The history of technology in the classroom gives one pause. Too often, inflated promises have been followed by a burst of enthusiasm and partial implementation and then by discouragement and despair, broken morale and broken machines (Tyack & Hansot, 1985, p. 35).

There was a time when electricity was a novelty; something to be harnessed, to be understood. With time, not only did mankind harness electricity and understand it, but electricity became a mainstay of modern society. Today we cannot imagine our lives without light bulbs, refrigerators, and all the wonderful appliances that run on electricity. Similarly, there was a time when the telephone was a toy for the wealthy in their homes. Only the very well off could afford one of those "contraptions." Why would the hoi polloi have a need for a telephone at home? Again, with time, not only did the hoi polloi have a need for the telephone in the home, but phones became self contained (i.e., didn't need an operator as a mediator), smaller, and portable with the advent of the modern 21st century cell phone. But cell phones too rely on electricity; and as with electricity, we cannot imagine our lives without the telephone in all its forms.

Electricity was a major force in the development of the modern computer. Today's modern computer has a long and distinguished history, but less for microcomputers in classrooms in general and specifically in the School District of Philadelphia. This chapter is divided into three parts. The first section will provide a brief history of computing from early efforts at calculating machines, to ENIAC in the 1940s through the sophisticated machines and wireless technology used in the early 21st century. The second section will provide a brief discussion of computers in schools, and the last section will provide an introduction to computers in the Philadelphia schools.

A Brief History of Computers

While some take the history of computing all the way back to the abacus (Sharp, 2005, p. 3), or to Blaise Pascal's calculating machine from 1642 (Provenzo, Brett, & McCloskey, 1999), this book will begin with the Difference Engine and Analytical Engine developed by Charles Babbage in 1835 (Spencer, 1992a). Selected computing milestones will be presented, leading to the powerful microcomputers used today.

Charles Babbage began a critical examination of logarithmic tables used to make calculations accurate while at Trinity College, Cambridge, in 1810. He dreamed of a machine that could do these types of calculations, thus making them easier to complete. In 1822, Babbage designed what he called the Difference Engine, considered to be the first automatic calculating machine, specifically constructed for preparing mathematical tables. His second invention was the Analytical Engine, the first universal automatic calculator. This machine had all the components of a modern general purpose digital computer with memory (1,000 words of 50 digits each), control (via sequences of punched cards), arithmetic unit (addition and subtraction done in one second), input (via punch cards), and output (via punch cards or printed copy). Unfortunately, both of Babbage's machines were mechanically flawed (Provenzo, et al., 1999, p. 25). He was 100 years ahead of his time in terms of sophistication of technology available to him. In any event, Babbage is considered the grandfather of modern computing.

Herman Hollerith took a job with the U.S. Census Bureau in 1879. During the next few years he became preoccupied with trying to solve the mammoth "problem of automating the tabulation work of the census" (Spencer, 1992b, p. 490). His Tabulating Machine, developed after 1884, was an amazing machine that allowed sensing of holes through electrical contacts, which then allowed the cards to be easily sorted into pockets in a sorting box, then to be sorted by hand. With Hollerith's Tabulating

Machine, he invented the modern punch card (Austrian, 1982). In 1896, Hollerith formed the Tabulating Machine Company, which he sold to become part of the Computing-Tabulating-Recording (C-T-R) Company. In 1924, the C-T-R Company was renamed the International Business Machines (IBM®) Company. Hollerith is credited with automating data processing (Spencer, 1992b).

John Atanasoff was determined to create a machine that would allow him to do calculations faster than he could do. In 1941, Atanasoff, with Clifford Berry, a graduate student in engineering, built the first electronic digital computer, the Atanasoff-Berry Computer, or the ABC. It used binary logic circuitry and had regenerative memory. Their work came to a halt thanks to World War II, the ABC was never patented, and their work was largely ignored (Mollenhoff, 1988).

Howard H. Aiken dreamed of developing a large scale calculator, and began work on his dream with engineers at the IBM lab in Endicott, New York in 1939. IBM, in 1944, completed the Mark I (Cohen, 1999; Lee, 2000). The Mark I was 51 feet long, eight feet high, two feet thick, and weighed five tons. It was extremely noisy but could do three calculations per second (zipping along) using data from punch cards. The Mark series of computers was the forerunner of the IBM360 series that was responsible for making IBM a giant in computer technology.

John Mauchly, who taught at the Moore School of Engineering at the University of Pennsylvania, learned about the work of Atanasoff and used the designs on another computer. After the outbreak of World War II, Mauchly took a training course in electronics, designed for working with the defense industry, at the Moore School of Engineering at the University of Pennsylvania, where he later taught (Spencer, 1992c). Working together, Mauchly and J. Presper Eckert developed an operational electronic digital computer, finished in 1946, that they called the Electronic Numerical Integrator and Calculator, or ENIAC (Goldschmidt & Akera, 2003). Like the Mark I, ENIAC was huge, weighing in at 30 tons. Eighteen thousand vacuum tubes conducted electricity generating tremendous heat, and special cooling units were needed to prevent overheating which would lead to a shutdown. This computer solved mathematical problems 500 times faster than previous computers, but it had a small memory and difficulty shifting from one program to another (Sharp, 2005, p. 8).

Other Milestones

Computers continued to develop and be adopted in the business arenas during the next three decades. During these same years, mainframe computer systems were developed to support education, most of which were based on the research by B.F. Skinner, discussed in a later section of this chapter. What is most important for this book is the development of the personal computer. In 1974 a company in New Mexico called MITS invented a computer that could be sold as a kit for people to assemble at home. The Altair 8800 is considered to be the first personal computer (Provenzo et al., 1999, p. 30). It was difficult to use and was soon replaced by the Apple computer.

Stephan Wozniak designed a personal computer in his parents' living room. His friend, Steven Jobs, suggested selling the computers (Spencer, 1992d). They called their venture "Apple®" to symbolize friendly and easy to use technology. The first Apple was sold to the public in 1976. The company was an instant success and models of computers evolved to the Apple II, Apple II+, Apple IIE, and Apple IIGS®. They introduced the Macintosh in 1984, which was a 32-bit[1] computer that came with a black and white nine-inch monitor and detached keyboard and mouse. The Macintosh was the first computer to have a graphical user interface (GUI), meaning you could navigate into and out of programs using a pointer to activate pictorial representations, or icons, on the screen. This was truly revolutionary for before GUI you had to know the operating system commands to open a computer application, which made operating a personal computer challenging for many people.

While all of this was happening, Paul Allen and Bill Gates were developing computer software. They created the first version of Microsoft BASIC in 1975, adapting BASIC programming language, developed in 1964 by John Kemeny and Thomas Kurtz, to run on the MITS Altair computer. Soon after, they partnered to create Micro-Soft., later to become Microsoft® (Spencer, 1992e).

IBM introduced the IBM PC® in 1981 using MS-DOS® as the operating system. Again a revolution in technology was taking place with the 16-bit IBM PC showing up the previous 8-bit computers as slow and obsolete. Like the original Apple computer, you had to know the operating system commands to access applications. The IBM PC had 16 kilobytes of memory, which was "heaps" at that time, and could be expanded to 256 kilobytes of memory. It came with two floppy disk drives, and used the Intel® 8086 processor for "lightning" speed. Over time, IBM introduced the IBM XT® model in 1983 and IBM AT® model in 1984, each with more memory and using the Intel® 80280 microprocessor which had 16 megabytes of memory and supported multitasking[2]. The IBM AT Model 2 came with one floppy drive and a hard drive (Asbrand & Bozmann, 1992).

In the mid-1980s IBM clones came on the market. These computers used the MS-DOS operating system and ran all the software supported on the IBM PC series. Popular clones were the Tandy® 1000 and 1200, Leading Edge, and Compaq® computers. These machines were less expensive than the IBM machines but did the same work. The next generation of computers arriving in 1985 and using the Intel386™ processor was produced by Compaq. The Compaq computer was a 32-bit computer at 16 MHz speed. The IBM PS/2 (Personal System 2) was introduced in 1987. The Intel486™ computer debuted in 1990 and ran at 32 MHz speed (Asbrand & Bozmann, 1992).

The Internet, another force in technology, was born in 1969 when the Department of Defense created ARPAnet for security and research purposes. ARPAnet was to connect defense researchers in academia and business through a centralized network that a nuclear war could not fully destroy. They developed protocols (rules for sending and receiving data) called the Internet protocol, or IP, to be used with a transmission control protocol, or TCP. These protocols allowed computers to "talk" to one another on the network. The advent of computer-mediated communication brought such inventions as electronic bulletin boards, inter-relay chat, electronic mail, and instant messaging, all of which have changed the way we communicate (Ceruzzi, 2003, p. 296).

What revolutionized access to the Internet was development of the World Wide Web, (i.e., the Web) by Tim Berners-Lee in 1989 at CERN (European Council for Nuclear Research) in Switzerland when he created hypertext mark up language (Quittner, 1999) and access to the Web through browsers in the year 1993. The first Web browser, Mosaic, was developed by Marc Andreesen and Eric Bina (Ceruzzi, 2003, p. 303). The Internet spawned the proliferation of Internet Service Providers (ISPs) like CompuServe® and AOL®. Before the Web and browser access to the Internet, ISPs allowed a subscriber to access text documents only, since images were not supported by the Internet. With the advent of the graphical interface Web browser, text, images, sound, and video became available to the desktop computer. Hypertext Mark Up Language (HTML) allowed a Web user to do on the Web what previously could only be done through Hypertext on the Macintosh computer. Using HTML hypertext links, a Web user could jump from Web page to Web page or from Web server to Web server in a matter of seconds just with a point and click of the mouse button. Now with wireless technology, the same access can happen anywhere/anytime through cell phones!

When the Web came into being, creating Web pages or Web sites could only be done by the truly technology savvy who took the time to learn HTML. Software developers soon added the ability for someone creating a word processing document to save the document as a Web page. Now *anyone* can create a Web page or Web site without having to know the language. This is akin to not having to know the operating system language to access software applications on a computer, making use of the computer for purposes not yet conceived easier.

The developers of the early software and computers could not have imagined what was happening in the computing world in 2006. Apple Computer's iMovies® software allows the average person to create movies that make the old reel-to-reel, 8mm or 16mm, movies of 1940s and 1950s look primitive. The Apple iPod®, initially intended to play digital music, is now being used for podcasting of music, videos, university lectures, and more to be accessed anytime and anywhere. All of this leads to pervasive computing.

Other Issues About Computers

In this brief history of the development of modern computers, several topics have not been discussed, but will be touched upon here. One obvious one is that *Time* magazine named the computer as the Machine of the Year in 1982. Instead of naming their normal person of the year, the editors of *Time* took stock of how computing was impacting every aspect business and life, predicting that computers would change the way people interact with each other and computers as well (Friedrich, 1983).

The history of software development is long and extensive. Initial software available was for mathematical computation, which stands to reason since computers were first developed to solve vast mathematical problems like calculating the census data, analyzing weather patterns, or breaking secret codes. One of the first software applications for the Apple computer introduced in 1979 for the Apple II computer was Visicalc, a spreadsheet application for business (Ceruzzi, 2003, p. 267). This program was the standard for spreadsheets until the mid-1980s when Lotus® 1-2-3 came on the scene and brought graphical capabilities integrated into the program. Microsoft's Excel® was their contribution to the mix of spreadsheet options. Similarly, Harvard Graphics® was an early program used in business for graphing and creating slideshow presentations. Compared to Microsoft PowerPoint® today, the early Harvard Graphics software represents the stone age of presentation software.

There were several word-processing software options starting with Bank Street Writer, WordStar®, Word Perfect®, and Microsoft Word. Over time Word Perfect and Microsoft Word survived. The first integrated set of software applications was AppleWorks®, which included word processing, spreadsheet, database, drawing, and paint options. The advantage of an integrated application was that you could move seamlessly between programs and easily import products from one program into another. While these actions are done easily today in the Microsoft Office Suite of programs, it was unique when AppleWorks was introduced.

Desktop publishing came onto the scene in the mid-1980s on the Macintosh. The ability to create brochures and pamphlets, among other products, where text and images were easily included on the screen in columns, or rotated for effect, was previously accomplished only in publishing offices by experts. Now anyone could "publish" newsletters, brochures, and pamphlets, to name a few. Desktop publishing introduced the concept of "what you see is what you get," or WYSIWYG, to the computer vernacular. The user was in control (Norvelle, 1992).

Multimedia also developed during the last 20 years of the 20th century. Systems could be created that networked a computer with a videodisk player for access to still and moving images, a microphone to record sound, and speakers to play recordings—a multimedia station was created. There were creative software applications developed to allow the user to create a slideshow in which there might be links to play a videodisk image or video clip or to play an audio recording created by the user. One such program was HyperStudio® (Holcomb, 1992).

In 1998 Apple Computers introduced the iMac® computer. It was a unique design for the time, with the CPU and monitor all in one piece. While the design was what captured many people's attention, this was the first personal computer without a disk drive other than a CD-ROM drive for loading software. This design made file sharing a challenge requiring either an external disk drive or file sharing through the Internet. The iMac was the first Mac to feature the Universal Serial Bus (USB) port (Michaels, 2001).

Summary

Provenzo (1986) compared how microcomputers have changed the way people do business and think about how they work to the impact of the Gutenberg printing press on publishing. Prior to the printing press, books were created by scribes and copy clerks. Each book was "written" by hand, meaning it took a long time for new copies of a book to be developed. Basically, this meant books were extremely expensive and considered precious. Only the wealthy owned books as valuable possessions, or they resided in university and monastery libraries only to be used in the library. The printing press turned that system upside down reducing the cost of developing copies of books, also eliminating the possibility of errors created by careless scribes or copiers. Not that in 1450 the average person was buying books because they were still expensive and most people could not read, but the printing press made it possible for books to be used anywhere at anytime by anyone.

Early computers were used only by engineers, physicists or mathematicians at universities, in the government, or companies working for the government. Users had to have knowledge of the language used by the computer. As computer technology became smaller and less expensive, the microcomputer was born making computers accessible to the average person rather than specialists. Now computers are found in most homes in the United States. According to the U.S. Census Bureau, "In 2003, 70 million American households, or 62 percent, had one or more computers, up from 56 percent in 2001. In 1984, the proportion of households with a computer was 8 percent" (U.S. Census Bureau, October, 2005, paragraph 3). They are a virtual requirement in business offices. Where businessmen once depended on the steno/typing pool to develop documents, now they create the electronic document themselves saving much time.

This portion of this chapter has given an overview of the history of modern computing, starting with Charles Babbage's Difference Machine. Even though this invention did not really work as intended, Babbage is considered the grandfather of modern computing. If you look at the rationale give by each inventor over time, each wanted to solve a mathematical problem with which they had struggled. The computer, basically a calculator, made the work easier. As before the Gutenberg printing press when only the "elite" had access to books, before the microcomputer only the "elite" had access to computers. Now microcomputers are ubiquitous.

The Internet and World Wide Web have further revolutionized how communication occurs around the world. Documents can be signed electronically and sent electronically as official documents of record. E-mail and the Web have opened up opportunities for global economies unheard of before. Thomas Friedman, in his book *The world is flat* (2005) indicates how this globalization of access to communication networks means work that used to need to be done only during normal United States daylight hours (e.g., 9 to 5 in EST, PST, etc.) is now done anytime and anywhere. It is important to remember that all of the modern development of computers has occurred in a span of 60 years. The World Wide Web came into being in 1993. The fast track of computer use in so many areas may have ultimate consequences inconceivable at this time. Only time will tell. In the mean time, microcomputers, the Internet, and the Web are with us as we hurtle headlong into the 21st century.

A Brief History of Computers in Education

Seymour Papert, at the beginning of *The children's machine: Rethinking school in the age of the computer* (1993), asked the reader to imagine time travelers from the nineteenth century coming to visit. The group would include surgeons and school teachers. Papert goes on to suggest that the surgeon would be totally confused when visiting a modern hospital, with all the technology, specialization, and scope. On the other hand, the school teacher would no doubt feel right at home with today's classrooms not much different than before. With more and more computers entering classrooms through efforts of federal technology grant opportunities for the most part, this prediction may not be as true in 2007 as it might have been in 1993; however, teaching styles have not changed much even with computers present in classrooms.

When computers were invented the problem being solved was a mathematical one, not a creative arts, literacy, or social studies problem. And indeed, the first uses of microcomputers in schools were in mathematics or teaching programming. However, over time teachers discovered more powerful uses for microcomputers. This section of Chapter I will present a brief history of microcomputers coming into schools in general and discuss the possible impact of microcomputers on teaching and learning.

A Brief History of Microcomputers in Schools

While some would classify the move of computers into schools as a revolution, it is perhaps more of an evolution. It is changing the nature of schooling as we have known it since the slate board came into urban classrooms in the 1830s (Mehlinger, 1995, p. 92). "From 1978 to about 1982, the inclusion of computers in schools was primarily a grass-roots movement, often led by a single teacher. This usually resulted in only one teacher in a building using computers" (Roberts, Carter, Friel, & Miller, 1988, p. 1). Not all teachers were enamored with computing. They were not user-friendly and there were little to no options for application to the curriculum. But some parents were seeing that the computer was the "way of the future" and wanted their children to be prepared for the future. Let's face it, could *Time* magazine (Friedrich, 1983) have been wrong? But, where did this start?

Programmed instruction and task analysis surfaced in the 1950s. "The origin of educational technology [dates] from the work of B.F. Skinner and others on programmed instruction" (Morgan, 1978, p. 143). Skinner's theory of behavioral reinforcement and his advocacy of its application to education initiated the programmed instruction movement. This movement can be characterized by clearly stated behavioral objectives, with small frames of instruction, self pacing, active learner response to questions, and immediate feedback. Moving from programmed texts to programmed instructional media was a natural evolution since computers are infinitely patient and objective, doing these functions best.

Norman Crowder took the process of programmed instruction to another level. Basically Skinner's process was linear—start at the beginning and take incremental steps to reach the end. Crowder's idea was to have the process branching so students would go to those parts of the instruction that they need as demonstrated through their responses to questions. He called his idea "intrinsic programming" since the student determined the path taken accounting for individual differences in past learning and learning abilities. Using Crowder's ideas, Gordon Pask created a link between programmed instruction and computer-assisted instruction. The computer was programmed to assess the student's style and rate of learning and then to customize the next steps in the learning process to go forward or remediate. Of course, this decision tree required predetermined options and was still based on the behavioral theory of learning (Niemiec & Walberg, 1989).

In terms of teaching and learning, early uses of microcomputers were direct descendants from uses on mainframe computers. In the 1960s computers were used in education for computer-assisted instruction, or CAI developed for the mainframe computers. To access these programs, schools had to enter into a time-sharing arrangement with a university that had a mainframe computer with the required software. These arrangements could be expensive (Roblyer, 1992, p. 335). CAI was intended to help students hone skills, not to advance knowledge through higher order thinking activities. CAI came into development at the same time as the programmed instruction movement that followed the "Skinnerian teaching machines and reflected a behaviorist orientation" (Saettler, 1990, p. 307).

While considered by many to be mostly a computer language, LOGO is considered by some to also be an example of CAI. LOGO was first imagined by a research team in the late 1960s at Bolt, Beranek, and Newman, Inc, a Cambridge, Massachusetts, consulting firm, and developed at the Artificial Intelligence Laboratory at Massachusetts Institute of Technology (Saettler, 1990). Researchers wanted to develop an environment where children could learn mathematics and computer programming by non-mathematical means (Watt, 1990). Seymour Papert played a major role in developing LOGO by using his theoretical approach based on Jean Piaget's concepts of children's thinking. Compared to the previously described CAI projects that were based on behaviorist theories, LOGO was in stark contrast. Papert's approach was from a constructivist theory of learning where learners construct their own knowledge about the world, test these ideas, and modify them as needed. A shift was beginning to take place away from behavioral theory of learning, which focused on external causes of changes in behavior or learning, to the cognitive theory of learning, which viewed learning as a constructive process with an actively engaged learner linking old and new knowledge (Clark & Sugrue, 1995).

LOGO was quite popular in schools in the early 1980s for several reasons. Papert's book *Mindstorms* (1980) suggested that children who learned LOGO would easily gain in understanding of mathematics, problem solving, and computer science without much direction from the teacher. Papert's approach to student-centered learning was in sync with the desire, at that time, to teach computer literacy by allowing students to control the computer through programming. Many schools embraced LOGO in the early 1980s, only to reconsider several years later. Papert's interpretation of Piaget's theories of children's thinking was not necessarily realistic. His approach was to leapfrog a developmental stage moving from preoperational (ages 2-7) to formal operations (ages 11-15) skipping the concrete operations stage (ages 7-11) completely (Cuban, 1986).

Of course CAI was not the only way microcomputers impacted classrooms. Initially there was little in the way of educational software. An initial concept was that teachers would develop their own software using BASIC as the programming language, but there were three main problems with this concept. First, if teachers had access to a microcomputer, which very few did, teachers did not know BASIC. Secondly, teachers were afraid they would be replaced by a computer. As with every new technology introduced into schools, there was the impression that the technology could do the job of teaching better than teachers giving teachers the notion that they would soon be replaced. No technology has been able to effectively replace all of what teachers accomplish for students (Cuban, 1986, 2001). And thirdly, teachers were concerned what they would do if the computer stopped working since they were not technicians and did not want to be. Part of the insecurity included a level of "not wanting to look foolish in front of students."

The basic question asked by teachers was, and is, "How would I use a computer in my classroom?" As noted above, early 1980s uses were limited to teaching programming languages such as BASIC and LOGO. This approach took the philosophy that students needed to know how the computer worked and that "computer literacy" was about being able to control the computer through the programming language. In fact, the concept of "computer literacy" was not uniformly defined, and often referred initially to being able to program in a computer language and evolved into use of productivity tools.[3] Teachers did not know programming languages, so this instruction was relegated to a computer teacher in a computer lab, if the school had one. By giving responsibility to the computer lab teacher to teach with and about computers, classroom teachers did not need to learn.

Productivity tools were next wave in the microcomputer revolution/evolution. The AppleWorks integrated suite of applications, introduced in 1984 for Apple computers, provided classrooms with word processing, spreadsheet, database, paint, and draw applications all integrated for seamless transfer between options. Suddenly, there was a move away from teaching programming. People needed to know how to word process and keep track of budgets and materials, and the paint and draw programs offered a variety of options.

It should also be pointed out that the early computers either had no hard drive or limited memory. The first computers were 8-bit, very limited by today's standards. As computing developed to 16-bit and 32-bit computers, computers were able to do so much more, giving them flexibility beyond the dreams of the early developers. The early machines used the 5 1/4-inch floppy disk that could store between 100 and 1,200 kilobytes of data. The 3 1/2-inch High Density micro diskette could hold 1.44 megabytes, or MB of data. Then came the Zip® disk technology (100 MB), the CD-ROM technology (650 MB), and USB flash drives (from 16 MB up to 8 GB capacity), among many others all of which revolutionized what could be done by extending the possibilities of file size and storage (Forcier & Descy, 2008). However, with expansion of computing power and storage capacity, Howard Becker points out that in "1992, only one-fifth of school computers had a hard disk drive for storing large programs and data files, and only one-fifth were connected to other computers through a local area network" (Becker, 1998, p. 21). While there were many computers in schools in 1992, they could not manage the latest software available, and they would not be capable of accessing the Internet when the World Wide Web was invented the following year. Development of computing power has surpassed the capacity and budgets available in schools time and again, leaving schools often with outdated computers because they could not afford to upgrade to newer models.

Robert P. Taylor (1980) proposed a set of classifications for educational applications: Tutor, Tool, or Tutee. In the *Tutor* role, the computer is the teacher. The assumption is that there are some teaching functions that a computer can do more effectively and efficiently than the teacher can. As a tutor, the computer presents information, questions are asked or a problem to solve given. The student responds, the computer evaluates the response and determines the next course of action—to move onto the next learning unit or go back to review. Examples of the computer as a tutor are drill and practice programs, tutorials, simulations, and games.

The *Tool* role of the computer is more functional through productivity tools. Using word processing software, they can type and edit papers and reports easier than by longhand or using a typewriter. Similarly, the computer can be used as a creative tool by using paint or draw programs to illustrate a story or poem.

The *Tutee* role is when the student programs the computer to do what he/she wants. Students must learn a computer language in order to communicate with the computer for the computer to do what the student wants. They learn to write commands and the syntax necessary for a particular language, whether the language is BASIC or LOGO in the 1980s, or HTML in the 1990s.

Cleborne D. Maddux (1986) expanded on Taylor's work. Maddux offered that software could be categorized as one of two types of applications. Type I applications use the computer to do things already being done in the curriculum. Examples of Type I uses included administrative, drill and practice, tutorials, CAI and assessment of learning. These uses were already in place even before microcomputers in classrooms. Type II applications allow new models of teaching and learning not conceived before. Examples of Type II uses are children as programmers (e.g., LOGO), writers using word processing, educational simulations of dangerous, expensive or unavailable experiences, and augmentative equipment to assist the handicapped (e.g., voice recognition, speech synthesis).

The technology revolution/evolution was just beginning. Laserdisc technology was available in schools by the mid-1980s. In the early 1990s, textbook companies were pairing textbooks with laserdiscs. Barcodes in the text used with a barcode reader would allow the teacher to access resources on the laserdisc any time. The materials were film footage, video clips, still images, and diagrams intended to help the teacher present the materials to the class. This example is clearly teacher focused. At the same time, HyperCard® was introduced with Macintosh computers. Students who learned HyperCard programming language could create projects that were non-linear and could incorporate images or links to a laserdisc to access specific information there for the project. This example is clearly student focused. The movement away from the Skinnerian view of behaviorism toward the philosophy of constructivism promoted by John Dewey, Jean Piaget, and Lev Vygotsky was implicit in these activities.

Today, wireless technologies abound and "high tech schools" are the rage in some parts of the United States. But, while the technophiles praise the wonders of computing through special journals and conferences, the actual impact on teaching and learning is not clear cut.

Impact of Computers on Teaching and Learning

In 1983, at the same time microcomputers were beginning to be seen in public schools in the United States, *A Nation at Risk* was published by the National Commission on Excellence in Education. One of the "Five New Basics" recommended to be included in high school graduation requirements was "computer science." The report indicated that high school graduates should understand computers as information, computation, and communication devices, be able to use the computer to study the other Basics (e.g., English, mathematics, science, and social studies) and the industry of computers and related technologies (Culp, Honey, & Mandinach, 2003). This gave school districts and schools license to move toward including microcomputers in high schools, at the very least. How this recommendation related to elementary or middle schools was something to be considered by school districts. With the recent No Child Left Behind Act (Department of Education, 2001), students are to be technologically literate by eighth grade. Technology literacy is defined as the ability to use computers to communicate, to locate and manage information, and to use these tools to support learning of English, science, mathematics and social studies; a reflection of what was proposed in *A Nation at Risk* nearly 20 years earlier.

How quickly did computers "invade" schools and classrooms? Surveys about computers in schools reported by Howard Becker (1985, 1991) and Anderson (1993) noted that during the 1980s and early 1990s, schools had relatively few computers and related equipment, particularly up-to-date equipment compared to number of potential users. By 1983, computers were being used for instruction in more than 40 percent of all elementary schools and more than 75 percent of secondary schools (Center for Social Organization of Schools, 1983). This same study found that, on average, microcomputers were used for only two to three hours a day. By 1985, a study at Johns Hopkins University reported that there were at least one million microcomputers in American elementary and secondary schools (Becker, 1986). According to the U.S. Congress, Office of Technology Assessment (OTA), the percentage of schools with one or more computers grew from nearly 18 percent in 1981 to 95 percent in 1987. (OTA, 1989) By 1995 the Office of Technology Assessment reported there were nearly 5.8 million computers in schools. And by 2003, 100 percent of all public schools in the United States had access to the Internet with 93 percent of all public school classrooms having Internet access. The ratio of students to computers with Internet access was 4.4 to 1, down from 12.1 to 1 as reported in 1998 (National Center for Education Statistics, 2005). These numbers will, no doubt, continue to evolve.

Student use of computers is another topic altogether. In 1986 Becker reported the average student used a computer approximately 30 minutes a week, while some used a computer far less. Drill and practice was the primary use in elementary schools with programming and computer literacy prevalent in high schools. In the ensuing 20 years, access to computers improved per student as demonstrated above, but actual time using computers changed little. The primary access to computers is still in computer labs requiring scheduling of students throughout the week. Depending on the size of the school and number of computer labs available, students may still have concentrated instruction in computing, and perhaps access to computers, only during computer lab time of approximately 45 minutes a week. In some schools, class scheduling into a computer lab is only to allow the classroom teacher to have a "prep" period[4] during the course of the day, which results in a lack of connection or coordination between what happens in the classroom and the computer lab.

Research on change in schools has demonstrated the importance of awareness of ways in which the organization of educational systems impact adoption, implementation and adaptation of innovations (Crandall & Loucks, 1983; Fullan, 2001; Gross, Giaquinta, & Bernstein, 1971; Oettinger, 1969; Sussman, 1977; Van den Berg, Van Welzen, Miles, Ekholm, & Hameyer, 1986) Or as noted by Sheingold, Kane, and Endreweit in 1983, "...the effects of microcomputers on education will depend, to a large extent, on the social and educational contexts within which they are embedded" (p. 431). They also predicted, six trends emerging from research studies on microcomputer use in schools that would impact use and effectiveness, which are still useful today. The six are:

1. Access to microcomputers,
2. Emergence of new roles in response to microcomputers,
3. Integration of microcomputers into elementary classrooms,
4. Quantity and quality of software,
5. Preparation of teachers for using microcomputers, and
6. Effects and outcomes of the instructional use of microcomputers.

Access to computers, as noted above, has improved over the last 25 years. However, there is still something considered to be the "digital divide." Modern, up-to-date computers are more likely to be found in affluent, suburban schools, while poor inner city and rural schools are more likely to have older models. This means that the inner city and rural schools' computers are less likely to have enough memory and power to run the most current versions of software, and may not be capable of accessing the Internet. This is not even addressing the infrastructure issues of older school buildings. Newer construction and the Telecommunications Act of 1996 that created the E-Rate afforded districts the opportunity to upgrade schools to ensure better infrastructure for computing (FCC, n.d.). Accessibility is still a key factor.

Integration of computers into elementary school classrooms was more of a challenge than into secondary schools. In secondary schools typing classes were in the business curriculum, so keyboarding was a natural next step. Programming in languages like BASIC and PASCAL were also prevalent, and seemed to be necessary through the 1980s to demonstrate that the students were "computer literate." For elementary schools, programming did not seem to be appropriate, until LOGO came along. Having software that was both engaging and educational for early ages was a great challenge.

Also as noted before, in 1983 there were very few software options for microcomputers. The software industry was being born. With the proliferation of educational software, a new challenge arose—identifying quality software that complemented the curriculum. The Micro SIFT project and Educational Products Information Exchange (EPIE) were developed for just this purpose (Sharp, 2005). Teachers still struggle with the challenge of identifying pedagogically appropriate software for their students, and some school districts have committees that identify software options that are congruent with the curricula at the different grade levels. The latest problem is identifying good quality educational Web materials in the same way as identifying good quality educational software.

Colleges of education, in the early 1980s, were not prepared to help teachers learn how to use computers in the classroom. There were no computer labs available. There were no computer teachers, just like there were no computer teachers in public schools. All of these would develop at the same time computers were infiltrating public school classrooms. Not an ideal situation. At the same time, school districts needed to help classroom teachers learn about microcomputers. But change was to be constant. No sooner did teachers learn how to use the Apple IIE than the Macintosh came forth. So many changes in so little time meant as soon as a teacher became familiar (not proficient) with one, new computers came out.

Lastly, Sheingold et al. (1983) noted the need for evidence that microcomputers would make a difference in teaching and learning. Niemiec and Walberg (1992) conducted a meta-analysis of the findings from 13 research syntheses on CAI conducted between 1975 and 1987 and found the average effect size[5] was .42. This indicated that the average student who received CAI instruction scored at the 66th percentile as compared to the control group distribution. Others have also conducted meta-analyses of the impact of CAI on student outcomes (Lipsey & Wilson, 1993) or the effects of computer programming on student outcomes (Liao & Bright, 1991). Overall, these studies have documented the positive effects of educational technology on student achievement. (Schacter, 2001; Sivin-Kachala, 1998; Wenglinsky,

1998) Unfortunately, these studies also look at different types of technology. Overall, these analyses have not provided guidance for how to appropriately integrate and use technology in schools and classrooms. In 1983 there was precious little known about how microcomputers could effect schools and classrooms. Sivin-Kachala and Bialo (1994) reported that in review of 133 research reviews and reports, computer technology in the 1980s generally has been credited with "motivating students, aiding instruction for special needs students, improving students' attitudes toward learning, and motivating teachers and freeing them from some routine instructional tasks, enabling them to better utilize their time" (p. 6). Keys to success would be a combination of responding to each of these six issues over time.

One important component of computers in classrooms that came out of the Apple Classrooms of Tomorrow (ACOT^{SM}) studies, among others, is that the answer is not the presence of the computers, but the integration of computers into the curriculum (Brunner, 1990; Roschelle, Pea, Hoadley, Gordin, & Means, 2000). As Mellon so delicately pointed out, if technology is in place and students do not learn, then the teacher is blamed; however, if technology is poor or in short numbers and students do not learn, then the lack of technology is to blame (Mellon, 1999). While this is simplistic, both arguments ignore the fact that the learner is part of the equation. If the learner is not motivated to learn, computer integration will not make him/her learn. Computers in classrooms are here to stay for the foreseeable future. Making sure they are used in pedagogically appropriate ways that are integrated into the curriculum, not added on or as a reward, will be key.

Glenn M. Kleiman (2000) presented five myths about computers in schools. The first myth is that by putting computers into schools there will be a direct correlation with improved learning gains and that more computers will result in greater gains in learning. However, without adequate professional development (on the uses of the computer and classroom management techniques to support use in the curriculum), technical support, availability of appropriate software, and integration of computers into the curriculum, adding more computers to a school will not result in improved learning gains. Maddux (1986) pointed to this tendency in schools saying "…if schools can only obtain a sufficient quantity of hardware and software, quality will take care of itself" (p. 2). Unfortunately, Maddux was wrong.

The second myth is that there are agreed-upon goals and "best practices" that define how computers should be used in K-12 classrooms. How computers are best used in classrooms depends on many things, including the student population and the goals of the curriculum. However, often addition of computers in schools is not coordinated with the curriculum so they may not be used productively. Like myth #1, putting computers into classrooms does not necessarily translate into meaningful use.

The third myth is that once teachers learn the basics of using a computer, they are ready to put the technology to effective use. The Apple Classrooms of Tomorrow project demonstrated that teachers need to move through five stages of evolution (see

Chapter III) before using technology in the classroom seamlessly (Fisher, Dwyer, & Yocum, 1996; Gooden, 1996) These changes can only take place over a number of years with adequate support within the curriculum and for the technology.

The fourth myth is that the typical school district technology plan is adequate for putting technology to effective use. The reality is that technology plans tend to make having technology the goal rather than integrating technology into curricular goals. "... To use technology effectively we must fully integrate it into school improvement plans, special education plans, curriculum plans, professional development plans, and all the other plans formulated by schools and districts" (Kleiman, p. 13). Until this happens, computers in classrooms may reflect what Larry Cuban found in his study of classrooms in Silicon Valley—that they sit mostly unused by teachers and students (Cuban, 2001).

Lastly, the fifth myth is that equity can be achieved by giving schools in poor neighborhoods the same student to computer ratio as in wealthier neighborhoods. While there is a digital divide between the haves and have-nots, putting computers in schools is just one step. As with the previous four myths, unless all aspects of the enterprise are included in the upgrading, just adding computers will not make a difference. The key to ensuring equity, effective use, "best practices" used, and improvement in student achievement is to have a shared vision of how computers impact education, prepare and support teachers, design a curriculum with computers integrated (not just tacked on), and a long term commitment to supporting students and teachers.

In 2000, Becker reported the results from the Teaching, Learning and Computing Survey (TLCS). In this paper, he takes on suggestions by Cuban (1986, 2000) that computers are incompatible with classroom teaching requirements, and that teachers continue to not use them for student work during class. Becker reported that, from a nationally representative survey of 4th through 12th grade teachers, three conditions seem to predict use. While the majority of teachers do not use computers routinely during classroom time, due to factors related to the demands of the curriculum, lack of experience with computers, and lack of knowledge about how to effectively integrate computer options into the curriculum, the TLCS data suggests teachers are more likely to have students use computers actively and regularly during class time when they are a constructivist-oriented teacher with enough computers in their classrooms (i.e., clusters of five or more) and with a reasonable level of experience and skill in using computers themselves (Becker, 2000).

Summary

While CAI was the primary use of computers from the inception of the programmed instruction movement, CAI had little impact on education. Evaluation of the impact of technology on education is too complicated and complex to be assessed through traditional research designs. Mandinach and Cline (1997) noted that these designs are inadequate, inappropriate and simplistic in merely asking if technology works. The impact of technology must be considered in the context of student learning and motivation; interactions between students, teachers and computers; and the overall organization and climate of the school. To further complicate the process, technology and policies are ever changing and evolving. Between federal mandates, state mandates, district expectations and local school practices, isolating which factor is the cause of which effect is more than difficult. Teaching with computers in pedagogically appropriate ways by grade level and student needs will lead to greater understanding and ownership of knowledge. That belief is what drives school districts and schools to continue to pursue technology resources.

A Brief History of Computing in Philadelphia Schools

The Office of Educational Technology in the School District of Philadelphia has a rich history over the last 30 years or so. As chronicled in other cities, microcomputers came into Philadelphia schools serendipitously. There was no systematic plan or overall vision for microcomputers in schools in the late 1970s, as was true across the United States. Few people knew what to do with computers in schools and they were often seen as toys, not learning tools. At some schools (exactly how many is impossible to know), there were individual teachers who had purchased a computer on their own to play with and learn because they were curious. At other schools the Home and School Association members (mostly from business, who had microcomputers in their offices and saw the usefulness of them) thought computers were the future, so they raised money to put a few computers in the school library (B. McGee, 2006, personal communication, February 7, 2006). In either case, it would take one teacher who saw the potential of computers to try them out in his/her classroom (Mehlinger, 1995). At the same time, if the principal of a school had a vision of how these new tools could support the curriculum and teachers to support student achievement, teachers were more likely to be given what was needed to make microcomputers work in that school (A. Kramer, personal communication, March 3, 2006).

Initially, due to the paucity of educational software, programming was the major focus of microcomputers in classrooms. Being able to program the computer to do something (e.g., the microcomputer as the tutee and the student as the tutor, a la Taylor, 1980) provided opportunities for students to problem solve when their program did not run as expected. Problem solving and critical thinking skills were considered important learning goals in teaching programming. Programming continued to be a major focus in the School District of Philadelphia throughout the 1980s (M. Bamash, personal communication, March 10, 2006). There was a district wide LOGO user support group, called the Technology User Group, that met regularly to share ideas and projects, and to share ideas about new versions of LOGO as they arrived on the scene. There was the LOGO Day Conference for students where schools were invited to bring students to an elementary school in northwest Philadelphia. The day was patterned after professional conferences such that students signed up for sessions to learn a skill, and moved between sessions throughout the day. They learned more about LOGO, among other programming options, over the years the student conference took place.

But this brief synopsis does not do justice to the whole story of microcomputers and Philadelphia schools. While it is not possible to chronicle the stories and experiences of all the teachers in the school district who made technology a central focus and helped establish the vision of micro computing in the district today, the examples below will provide evidence of how computing spread in the School District of Philadelphia slowly but deliberately. These are only a few examples. And it is important to recognize the importance of principals in this process. Without the principal supporting technology, technology did not play a major role (B. McGee, personal communication, February 7, 2006, A. Kramer, personal communication, March 3, 2006).

Dr. Sylvia Charp was credited as the individual who started the educational technology movement in the district in the late 1960s in the Division of Instructional Systems with an Instructional Computing Center (G. Brookstein, personal communication, February 28, 2006). Indeed, Dr. Charp was influential on an international scale, being the Past President of the American Federation of Information Processing Societies (AFIRS), U.S. Representative to the International Federation of Information Processing (IFIP), and Past President of the International Society for Technology in Education (ISTE®) (HEIA, 1998). She started her career in Philadelphia, and is credited with moving technology forward in Philadelphia, making Philadelphia one of the leaders in educational technology in the early days of educational computing (J. Snyder, personal communication, March 3, 2006). Dr. Charp went on to be the founding editor of *T.H.E. Journal*. One way she moved computing forward in Philadelphia was through a million-dollar federal grant to develop a computer program for teaching biology and reading, called Project GROW, in concert with the Philco-Ford Corporation. The program was very high-end technology, but it did not implement well in the classroom (J. Snyder, personal communication, March 3, 2006). While perhaps not a positive outcome, such were the trials and tribulations of early educational computer program development.

From her early days with mainframe computers, the microcomputer revolution in Philadelphia was continued by Arlene Kramer in 1983, along with a number of talented individuals, including Jerry Snyder, Richard DiCaprio, Mel Levin, Cheryl Heath, Arnold Indictor, and Paul Hampel, among many others (A. Kramer, personal communication, March 3, 2006). Ms. Kramer taught mathematics at a high school in Northeast Philadelphia in the 1960s with one computer terminal in the school. She had a vision of how computing could make a difference in students' learning. There were no criteria in the early years of microcomputers in schools for who could be a computer teacher. Working in the Office of Computer Science and Technology, she and Mr. Snyder established three things—a basic computer curriculum by grade level, criteria for someone to apply to be a computer teacher and an assessment process to become a computer teacher (A. Kramer, personal communication, March 3, 2006, J. Snyder, personal communication, March 3, 2006). Before there was a technology curriculum and standards for teachers of computer science, teachers taught what they knew and what the available materials supported. The criteria and process for becoming a computer teacher survived for many years until the Commonwealth of Pennsylvania established criteria for certification in the early 21st century.

In the early 1980s, many different makes of computers were cropping up in schools—the Atari, Commadore, TRS80, Apple, among others. In order to best support those computers and the teachers, Ms. Kramer believed that the district should standardize on a single platform. Given all the options that were available at that point in time, Ms. Kramer believed that the Apple II computers should be adopted as the platform of choice as they were noted to be superior in quality and in the availability of good educational software (A. Kramer, personal communication, March 3, 2006).

Believing the principal to be a major factor in facilitating computer integration in schools, Ms. Kramer and Mr. Snyder wrote a handbook for principals on computers in the classroom. This book included information on copyright laws, the one-computer classroom versus the computer labs, and information on troubleshooting technology problems, among other things (A. Kramer, personal communication, March 3, 2006). In order to get principals to see the value of technology, the Office of Computer Science and Technology (later renamed the Office of Educational Technology) established a series of training programs for principals. Apple computers donated a Macintosh computer to each principal who completed all the training sessions. Ms. Kramer noted they knew the principal was key to the success of these programs (A. Kramer, personal communication, March 3, 2006).

Mr. Jerry Snyder was first assistant to Ms. Kramer in the Office of Computer Science and Technology. Initially, he noted, the emphasis on computer training was on programming in BASIC and LOGO (J. Snyder, personal communication, March 3, 2006). Most notably, Mr. Snyder was credited by Ms. Kramer for bringing Internet access into the district and into schools, along with laying the groundwork for the establishment of the district's wide-area-network. This was before the E-Rate opportunities of the late 1990s and early 21st century, so expanding access was a challenge. His efforts set the stage for what is now available in the district (A. Kramer, personal communication, March 3, 2006). Working with Ms. Kramer, several initiatives were started. Mr. Snyder and Ms. Kramer envisioned support for teachers in their schools through a teacher leader position. The Technology Teacher Leader (TTL) position was finally implemented in the 1990s. They emphasized training of teachers and support in the schools, including a train-the-trainer model. Their vision played out in the systems developing in 2006 (A. Kramer, personal communication, March 3, 2006).

The schools where computing took off first were high schools, and in particular in the vocational or business education programs (L. Bell, personal communication, March 2, 2006). First the computer labs were used for learning to type or keyboarding, which over time replaced the typewriters. In the early 1980s, word processing systems, like the WANG system, were used extensively in business word processing centers. Next came word processing software for the microcomputer, like the Bank Street Writer. While word processing systems had been in use for years as well as simple text editor programs, Bank Street Writer, and later AppleWorks, introduced options that were amazing to the user, like move blocks of text, or find and replace character strings. These are common place today, but not in 1983 and certainly not done on a manual or electric typewriter at that time. Here is where the Provenzo (1986) comparison is most poignant. Before systems like Bank Street Writer or AppleWorks, one needed to know the language of word processing, like the scribes in the 1400s. All of a sudden anyone could manage the process of putting the written word onto paper electronically—you did not need to know the command language to get it done. Computer science was also offered in many senior high schools during this period of time. The "beginners" course was a course in programming in PASCAL, which was the mandated course of study at this level starting in 1983, although many high schools continued to teach programming in BASIC as well.

The process was much different in the elementary and middle grade schools. What is fortunate is that there were many "revolutionary" teachers in the Philadelphia schools. Unfortunately, this book cannot relate all of their stories, but can celebrate them for what they accomplished and contributed to the School District of Philadelphia. A very few teachers will be chronicled to provide evidence of what these pioneers did.

The start of the Philadelphia technology story is confirmed by several former teachers consulted for this book. Mrs. Susan Cook was teaching fourth grade at an elementary school in southwest Philadelphia when she was asked by her principal to take a training workshop on programming in BASIC. She came back excited about what these new machines could bring to the classroom. In 1986 her school opened the computer lab with about 20 Apple II computers. In addition to LOGO programming options, the school district had partnered with MECC to license any of the MECC computer programs for use in all schools (S. Cook, personal communication, February 23, 2006). As noted above, her principal was totally supportive of her work. Then there was a change in leadership in the school. According to Ms. Cook, the new principal did not have the same vision of the role of technology in the school. Her efforts to provide a cohesive set of experiences for students were thwarted by making the computer class a prep class, rather than a required subject. Some grades never were scheduled into the computer lab, and others only once a week for 45 minutes. Ms. Cook went on to become an Educational Technologist in the south region of the district and then a member of the Educational Technology Group.

Another example of the serendipity of computers into schools is the story of Mr. John DiLello. Mr. DiLello's father was an assistant principal in a school in Philadelphia in the early 1980s (J. DiLello, personal communication, February 14, 2006). There was a policy, or practice in some schools in those days, that microcomputers were not to be left in schools over the summer, for fear of loss or damage. His father brought home an Apple II from his school for the summer. Mr. DiLello had never seen one before, but was willing to learn how to use it. The real epiphany was after Mr. DiLello became an English teacher in the late 1980s. Word processing software had arrived for the computer. He found that word processing software helped his own writing, and thought the editing process could empower his students to write better. He convinced his principal that he needed three or four computers to help his students write better and the principal made it happen. He became identified as the "computer guy" in the school, or the person to go to with questions about computers.

Mr. Mel Levin taught sixth grade social studies at a school in northwest Philadelphia in the early 1980s (M. Levin, personal communication, February 23, 2006). The Home and School Association installed four Apple IIE computers in the library. Mr. Levin noticed that no one was using them. He was always looking for new ways to engage his students, so he asked the principal if he could take one into his classroom. He developed ways for his students to program in BASIC, and then LOGO to develop critical thinking and problem solving skills, and collaborative projects. The principal noticed that Mr. Levin's students were very engaged in these activities and decided to convert the art lab into a computer lab using grant funds. Mr. Levin became the computer teacher in the school (M. Levin, personal communication, February 23, 2006). Without the support of the principal, this might never have happened.

Mr. Levin discovered that there were other teachers in the district who were using LOGO as a learning tool and started the Technology Users Group. These teachers would get together to discuss what they were doing and trade projects and the like. The group decided they wanted to expand the knowledge and use of LOGO to other students in the district. The principal at his school allowed them to try out a student conference on computing based on professional development conferences they had attended. The first Computer Conference for Students was in 1984. There were workshops on different uses of LOGO. Over the 12 years of the project between 500 and 600 students participated each year (M. Levin, personal communication, February 23, 2006). All of this effort was supported by his principal and Ms. Kramer in the Office of Computer Science and Technology (A. Kramer, personal communication, March 3, 2006). Support by the principal is clearly a theme.

Mr. Levin was named Educator of the Year in 1994 by *Electronic Learning* (Siegel, 1994). His development of the Technology Users Group was mentioned along with the computing conference for students. Mr. Levin was also mentioned in Mehlinger's book in 1995 as one of the technology revolutionists. "Action by state and federal governments and by business and industry has helped fuel the revolution, but they did not provide the spark" (Mehlinger, p. 118). Mehlinger credits the teacher technology revolutionists, like Mr. Levin, who sparked and nurtured the interest in using microcomputers by other teachers, as key. There were other technology revolutionists in the School District of Philadelphia, but they did not earn the same notoriety as Mr. Levin.

A former special education teacher, Mr. Vince DeTolla became a force within the school district because of his sincere belief and vision about how microcomputers could impact children's learning. He became first acquainted with microcomputers at a disciplinary school. There was a computer lab in the early 1980s of Commodore VIC20 computers. The students he worked with were challenged academically and behaviorally. When he saw that their interactions with the computer changed their behaviors and engagement, he knew they were onto something. MECC offered software for the Apple computers and Commodore. He and others looked at a lot of programs around reading and problem solving skills, in particular because these children had very poor reading skills (V. DeTolla, personal communication, March 6, 2006). He saw this as a movement, so he earned a masters degree in computing from Temple University, after which he moved to an elementary school in Center City Philadelphia to be the computer teacher. He was a LOGO teacher, and early adopter of Lego® Robotics. From his perspective, students who used LOGO improved in mathematics and problem solving skills.

His success at this school did not go unnoticed. He was recruited to a middle school in south Philadelphia where he became computer science teacher for the mentally gifted students. He noted, because mentally gifted students in Pennsylvania are categorized as special education students, he could basically write his own curriculum. So, these students did a number of collaborative projects using simulations and other new computer options to expand their knowledge and skills. At the middle school, to get other teachers to use computers in their practice, they tried to identify the "killer ap," which initially was word processing. All students needed to know how to word process, so they set up centers of six computers in the back of classrooms. Students could work in groups or individually on writing projects. This seemed to work for language arts (V. DeTolla, personal communication, March 6, 2006).

Mr. DeTolla became the Executive Director of the Educational Technology Group in the early 21st century, the office of instructional technology for the district. In discussing how technology moved into the district, he noted how they tried to get a computer in classrooms for teachers to use, but they did not follow up with deep professional development. Teachers would use the computer initially, but then it became the reward for students for good behavior. Over time the computers were relegated to the corner and abandoned. The early adopters were those who learned on their own. His portrait of this process is not unique to Philadelphia (Cuban, 1986, 2001).

Summary

As one can see from these former teacher stories, computing began in Philadelphia schools in a random fashion. Teachers who were intrigued by microcomputers and what they could do were the first to adopt, and often became the computer teachers in the schools. Programming was the first wave of microcomputer use to learn how to "teach" the computer how to do things, and to teach the students problem solving and critical thinking skills. Next, word processing software was seen as a modern tool for business education, and as a way to empower students in the writing process. However, it would not be until the Telecommunications Act of 1996 and the establishment of the E-Rate program for all schools in Philadelphia to gain in computing options. Overall, without a supportive principal, none of these efforts were effective beyond individual classrooms. And yet, the overall support for teachers in classrooms to sustain use after learning new skills was not yet fully in place. One deviation from this pattern was the Continuous Practice Improvement program the basis of which was ongoing support back in classrooms.

Overall, computers appear to be in schools to stay. Helping teachers use them appropriately for student learning appears to be a challenge well into the 21st century. With this background knowledge, this book will now present the theoretical lenses used in understanding the impact of the Continuous Practice Improvement professional development program.

References

Anderson, R.E. (1993). *Computers in American schools, 1992: An overview*. From *The IEA Computers in Education Study*. Minneapolis, MN: University of Minnesota.

Asbrand, D., & Bozmann, W.C. (1992). Microcomputers (personal computers). In G.G. Bitter (Ed), *Macmillan encyclopedia of computers*. New York: Macmillan, pp. 667-674.

Austrian, G.D. (1982). *Herman Hollerith: Forgotten giant of information processing*. New York: Columbia University Press.

Becker, H.J. (1985). *How schools use microcomputers: Summary of the 1983 national survey.* Second national survey of instructional uses of school computers. Baltimore: The Johns Hopkins University.

Becker, H.J. (1986). *Instructional uses of school computers.* Reports from the 1985 National Survey, nos. 1-3. Baltimore: The Johns Hopkins University, Center for the Social Organization of Schools.

Becker, H.J. (1991). How computers are used in United States schools: Basic data from the 1989 IEA computers in education survey. *Journal of Educational Computing Research, 7,* 385-406.

Becker, H.J. (1998). Running to catch a moving train: Schools and information technologies. *Theory into Practice, 37*(1), 20-30.

Becker, H.J. (2000). Findings from the teaching, learning, and computing survey: Is Larry Cuban right? Retrieved February 27, 2006 from http://www.crito.uci.edu/tlc/findings/ccsso.pdf

Brunner, C. (1990). What it really means to "integrate" technology. *Technology & Learning, 11*(3), 12, 14.

Center for Social Organization of Schools. (1983). *School uses of microcomputers: Reports from a national survey. (Issue no. 1).* Baltimore: The Johns Hopkins University, Center for Social Organization of Schools.

Ceruzzi, P.E. (2003). *A history of modern computing.* Second Edition. Cambridge, MA: The MIT Press.

Clark, R.E., & Sugrue, B.M. (1995). Research on instructional media, 1978-1988. In G.J. Anglin, *Instructional technology: Past, present and future.* Second edition. Englewood, CO: Libraries Unlimited, pp. 348-364.

Cohen, I.B. (1999). *Howard Aiken: Portrait of a computer pioneer.* Cambridge, MA: The MIT Press.

Crandall, D.P., & Loucks, S.F. (1983). *People, policies, and practices: Examining the chain of school improvement.* Vol. 10. Andover, MA: The Network.

Cuban, L. (1986). *Teachers and machines: The classroom user of technology since 1920.* New York: Teachers College Press.

Cuban, L. (2000, January). So much high-tech money invested, so little use and change in practice: How come? Paper prepared for the Council of Chief State School Officers' annual Technology Leadership Conference. Washington, D.C.

Cuban, L. (2001). *Oversold and underused: Computers in the classroom.* Cambridge, MA: Harvard University Press.

Culp, K M., Honey, M., & Mandinach, E. (2003). *A retrospective on twenty years of educational technology policy.* Washington, D.C., U.S. Department of Education, Office of Educational Technology. Retrieved on March 3, 2006 from http://www.nationaledtechplan.org/participate/20years.pdf

Federal Communications Commission. (n.d.). *E-Rate*. Retrieved February 27, 2006, from http://www.fcc.gov/learnnet/

Fisher, C., Dwyer, D.C., & Yocam, K. (1996). *Education and technology: Reflections on computing in classrooms*. San Francisco: Jossey-Bass.

Forcier, R.C., & Descy, D.E. (2008). *The computer as an educational tool*. Fifth edition. Upper Saddle River, NJ: Merrill Prentice Hall.

Friedman, T. L. (2005). *The world is flat: A brief history of the twenty-first century*. New York: Farrar, Straus, and Giroux.

Friedrich, O. (1983, January 3). Machine of the year: The computer moves in. *Time Magazine, 121*(1). Retrieved March 5, 2006 from http://ei.cs.vt.edu/~history/Time.MOTY.1982.html

Fullan, M. (2001). *The new meaning of educational change*. Third Edition. New York: Teachers College Press.

Goldschmidt, A., & Akera, A. (2003). John. W. Mauchly and the development of the ENIAC computer. Retrieved February 17, 2006 from http://www.library.upenn.edu/exhibits/rbm/mauchly/jwmintro.html

Gross, N., Giaquinta, J., & Bernstein, M. (1971). *Implementing organizational innovations: A sociological analysis of planned educational change*. New York: Basic Books.

Higher Education in the Information Age. (1998). Dr. Sylvia Charp. Retrieved on April 12, 2006 from http://www.upenn.edu/heia/people/bio/charp.html

Holcomb, T.L. (1992). Multimedia. In G.G. Bitter (Ed), *Macmillan encyclopedia of computers*. New York: Macmillan, pp. 683-691.

Kleiman, G.M. (2000). Myths and realities about technology in K-12 schools. In D.T. Gordon (Ed.), *The digital classroom: How technology is changing the way we teach and learn.* Cambridge, MA: The Harvard Education Letter, pp. 7-15.

Lee, J.A.N. (2000). Howard Aiken's third machine: The Harvard Mark III calculator or Aiken-Dahlgren electronic calculator. *IEEE Annals of the History of Computing, 22*(2), 62-81.

Liao, Y.K., & Bright, G.W. (1991). Effects of computer programming on cognitive outcomes. A meta-analysis. *Journal of Research on Computing in Education, 24*, 367-380.

Lipsey, N.W., & Wilson, D.B. (1993). *Practical meta-analysis*. Thousand Oaks, CA: Sage.

Maddux, C.D. (1986). Issues and concerns in special education microcomputing. *Computers in the Schools, 3*(3-4), 1-19.

Mandinach, E.B., & Cline, H.F. (1997, March). Methodological implications for examining the impact of technology-based innovations: The corruption of a research design. A paper presented at the annual meeting of the American Educational Research Association.

Mehlinger, H.D. (1995). *School reform in the information age*. Bloomington, IN: Center for Excellence in Education.

Mellon, C.A. (1999). Technology and the great pendulum of education. *Journal of Research on Computing in Education, 32*(1), 28-36.

Michaels, P. (2001). The iMac turns three. Retrieved on April 6, 2006 from http://www.macworld.com/2001/08/bc/16imac_3years/

Mollenhoff, C.R. (1988). *Atanasoff, forgotten father of the computer*. Ames, IA: ISU Press.

Morgan, R.M. (1978). Educational technology – adolescence to adulthood. *Education Communication and Technology Journal,* 26, 142-152.

National Center for Education Statistics. (2005). Internet access in U.S. public schools and classrooms: 1994-2003. Retrieved on February 27, 2006 from http://nces.ed.gov/pubsearch/pubsinfo.asp?pubid=2005015

National Commission on Excellence in Education. (1983, April). *A nation at risk*. Retrieved on March 5, 2006 from http://www.ed.gov/pubs/NatAtRisk/risk.html

Niemiec, R.P., & Walberg, H.J. (1989). From teaching machines to microcomputers: Some milestones in the history of computer-based instruction. *Journal of Research on Computing in Education, 22*(3), 263-276.

Niemiec, R.P., & Walberg, H.J. (1992). The effects of computers on learning. *International Journal of Educational Research, 17*, 99-108.

Norvelle, R.L. (1992). Desktop publishing. In G.G. Bitter (Ed), *Macmillan encyclopedia of computers*. New York: Macmillan, pp. 293-300.

Office of Technology Assessment. (1989). *Power ON! New tools for teaching and learning*. Lancaster, PA: Technomic Publishing Co.

Office of Technology Assessment. (1995). *Teachers & technology: Making the connection.* Report No. OTA-HER-616). Washington, DC: U.S. Government Printing Office.

Ottinger, A.B. (1969). *Run computer run: The mythology of educational innovation*. New York: Collier.

Papert, S. (1980). *Mindstorms: Children, computers and powerful ideas*. New York: Basic Books.

Papert, S. (1993). *The children's machine: Rethinking school in the age of the computer*. New York: The Free Press.

Polit, D.F., & Hungler, B.P. (1995). *Nursing research: Principles and methods*. Philadelphia: Lippincott.

Provenzo, E.F., Brett, A., & McCloskey, G.N. (1999). *Computers, curriculum, and cultural change: An introduction for teachers*. Mahwah, NJ: Lawrence Erlbaum Associates.

Provenzo, E.F. (1986). *Beyond the Gutenberg galaxy: Microcomputers and the emergence of post-typographic culture*. New York: Teachers College Press.

Quittner, J. (1999). Time 100: Tim Berners-Lee. Retrieved on February 24, 2006 from http://www.time.com/time/time100/scientist/profile/bernerslee.html

Roberts, N., Carter, R.C., Friel, S.N., & Miller, M.S. (1988). *Integrating computers into the elementary and middle school*. Englewood Cliffs, NJ: Prentice Hall.

Roblyer, M.D. (1992). Computers in education. In G.G. Bitter (Ed), *Macmillan encyclopedia of computers*. New York: Macmillan.

Roschelle, J.M., Pea, R.D., Hoadley, C.M., Gordin, D.N., & Means, B.M. (2000). Changing how and what children learn in school with computer-based technologies. *The Future of Children, 10*(2), 76-101.

Saettler, P. (1990). *The evolution of American educational technology*. Englewood, CO: Libraries Unlimited.

Schacter, J. (2001). *The impact of education technology on student achievement. What the most current research has to say*. Santa Monica, CA: Milken Exchange on Education Technology.

Sharp, V. (2005). *Computer education for teachers: Integrating technology into classroom teaching*. Fifth Edition. New York: McGraw-Hill.

Sheingold, K., Kane, J., & Endreweit, M. (1983). Microcomputer use in schools: Developing a research agenda. *Harvard Educational Review, 53*, 412-432.

Siegel, J. (1994). The energizer: Educator of the year, Mel Levin. *Electronic Learning, 14*(1), 28-30.

Sivin-Kachala, J. (1998). *Report on the effectiveness of technology in schools, 1990-1997*. Washington, DC: Software Publisher's Association.

Sivin-Kachala, J., & Bialo, E.R. (1994). *Report on the effectiveness of technology in schools, 1990-1994*. Washington, D.C.: Software Publisher's Association.

Spencer, D.D. (1992a). Charles Babbage. In G.G. Bitter (Ed), *Macmillan encyclopedia of computers*. New York: Macmillan, pp. 105-106.

Spencer, D.D. (1992b). Herman Hollerith. In G.G. Bitter (Ed), *Macmillan encyclopedia of computers*. New York: Macmillan, pp. 490-491.

Spencer, D.D. (1992c). John William Mauchly. In G.G. Bitter (Ed), *Macmillan encyclopedia of computers*. New York: Macmillan, pp. 648-649.

Spencer, D.D. (1992d). Wozniak, Stephan G. In G.G. Bitter (Ed), *Macmillan encyclopedia of computers*. New York: Macmillan, pp. 992-993.

Spencer, D.D. (1992e). Gates, William H (Bill). In G.G. Bitter (Ed), *Macmillan encyclopedia of computers*. New York: Macmillan, pp. 409-410.

Sussman, L. (1977). *Tales out of school: Implementing organizational change in elementary grades*. Philadelphia, PA: Temple University Press.

Taylor, R.P. (1980). *The computer in the school: Tutor, tool, tutee*. New York: Teachers College Press.

Tyack, D., & Hansot, E. (1985) Futures that never happened: Technology and the classroom. *Education Week, 4*(1), 44, 35.

U.S. Census Bureau. (2005, October). *Computer and Internet use in the United States: 2003: Special studies*. Washington, D.C.: U.S. Department of Commerce. Retrieved on March 3, 2006 from http://www.census.gov/prod/2005pubs/p23-208.pdf

U.S. Department of Education. (2001). *No Child Left Behind Act*. Washington, D.C.

Van den Berg, W., Van Walzen, W.G., Miles, M.B., Ekholm, M., & Hameyer, U. (1986). *Making school improvement work*. Heverlee, Belgium: Acco.

Watt, D.L. (1990). Logo. In G.G. Bitter (Ed), *Macmillan encyclopedia of computers*. New York: Macmillan.

Wenglinsky, H. (1998). *Does it compute? The relationship between educational technology and student achievement in mathematics*. Princeton, NJ: Educational Testing Service Policy Information Center.

Endnotes

1 A bit is the smallest amount of data handled by a computer. In binary code, a bit represents either a 1 or 0. Bits in groups of 8 are called bytes.

2 Multitasking means being able to have more than one software application open at a time, therefore doing more than one job at a time.

3 The International Society for Technology in Education (ISTE) took on this task along with the Department of Education, the National Council for the Accreditation of Teacher Education, and many of the learned societies to develop the National Educational Technology Standards for Students, Teachers and Administrators. While these are widely accepted, they have not been universally adopted in state or local standards.

4 Prep periods are used to allow "specialty teachers" to instruct in areas in which the regular classroom teacher might not be proficient. Art, music, physical education, and computer science are examples of "prep" subjects. They were

called "prep" subjects because they enabled the district to give elementary classroom teachers preparation time while the students worked with a "specialist" teacher. (A. Kramer, personal communication, March 3, 2006)

5 Effect size refers to "[a] statistical expression of the magnitude of the relationship between two variables or between two groups, with regard to some attribute of interest." (Polit & Hungler, p. 640)

Chapter II

Change Theory: A Model to Study Technology in Classrooms

One person claims that schools are being bombarded by change; another observes that there is nothing new under the sun. A policy maker charges that teachers are resistant to change; a teacher complains that administrators introduce change for their own self-aggrandizement and that they neither know what is needed nor understand the classroom (Fullan, 2001, p. 3).

How difficult is change? According to those who study change in schools and classrooms, change is something that is always occurring, or things remain the same. As Seymour Papert noted in 1993, if a teacher from the 18th century were to come into a classroom today, most likely they would feel right at home. In many cases students are still sitting in rows, perhaps not parallel rows, but rows nonetheless. There are books and papers around. Perhaps they would not be familiar with the overhead projector, but it would not take long to learn how to use it. Maps still hang on the walls. There might even be a computer somewhere in the room, but it might not be turned on, so no problem there. Making classrooms modern has not meant changing the typical look, feel, and layout of the traditional classroom.

"Constants amidst change... "(Cuban, 1986, p. 1). In education one can always count on change in curricula, practices, policies, procedures, management, and expectations, to name a few. Pressure for change comes from the top-down and bottom-up. And every innovation brought into schools is done with good intentions.

But change is difficult to achieve. And some teachers, who have seen new approaches to teaching come and go quickly over their years as teachers, take a jaundiced view as new ideas come into their schools. Just as they became knowledgeable and perhaps comfortable with the last approach, it is replaced with something quite different, so the learning begins again and the comfort-zone must be reestablished. So for teachers, in particular, the facilitators of change must not only show them how to use the innovation, but prove that it works and solves a problem that is agreed to be a problem.

Little (1993) described what she called five streams of reform. These included reform in:

1. Subject matter teaching (standards, curriculum change, and pedagogical methods)
2. Equity among diverse learners
3. The nature, extent and uses of student assessment, moving toward more authentic assessments
4. The social organization of schooling
5. The professionalization of teaching

Each of these streams has been considered innovations in education practice. This discussion of change in schools cannot be exhaustive for there are books and books devoted just to this topic. Therefore, five approaches to the study of change or reform in schools, especially relevant to the overall topic of this book, will be discussed.

C. Everett Rogers (1962, 1972, 1983, 1995, & 2003) was one of the first to address the processes of innovations becoming accepted into practice, or what he coined as the "diffusion of innovations." Rogers posited five stages of adoption:

1. Knowledge stage, or awareness of the innovation, but without complete information
2. Persuasion stage, or interest in the innovation, seeking more information
3. Decision stage, evaluation of the innovation mentally before actually trying it out, intending to try it out
4. Implementation stage, trial of the innovation to determine usefulness

5. Confirmation stage, or individual decides to fully integrate the innovation and promote it to others

He suggested for an innovation to be accepted into use, five criteria, or attributes, must be met. Rogers suggested that each of these was "somewhat interrelated empirically with the other four, but they are conceptually distinct" (2003, p. 223).

1. Relative advantage, or, is the innovation considered better than what is currently in use?
2. Compatibility, or, is the innovation compatible with the culture of the school?
3. Complexity, or, is the innovation simple and easy to understand?
4. Trialability, or, is the innovation available to be tested before adoption?
5. Observability, or, can the results of the innovation be observed by others?

Looking at the relative advantage of an innovation/change, the economic factors and rate of adoption need to be considered. "Initial cost of an innovation may affect its rate of adoption" (Rogers, 2003, p. 230). This is particularly important for schools and school districts that have limited funds. While an innovation technologically may be the best thing ever, the cost may prohibit schools using it until the price comes down, or schools can get a deal from the manufacturer. The innovation may give the district a level of prestige that may supersede the issue of cost, or the reverse could be true. Schools with computer labs full of Apple IIE computers would be considered extremely outdated in the year 2006, while schools with wireless laptop access for all teachers and students would be considered prestigious and in the forefront of the technology curve.

Compatibility with the culture of the school allows for compatibility with the sociocultural values and beliefs of the school (e.g., inclusive classrooms), previously new ideas (e.g., balanced literacy), and overall need for the innovation. Teachers may not know that they "need" an adoption until the change facilitator demonstrates to them what a great need they have. For instance, with the No Child Left Behind (NCLB) Act of 2001 came the need for schools to provide detailed evidence on how they were meeting the Adequate Yearly Progress (AYP), as defined by the law. Teachers had always assessed their students through classroom measures, but it became evident that these were not satisfactory for evidence for AYP. Regularly scheduled objective assessments (e.g., high stakes testing based on state or national subject specific standards) were required to provide evidence of student achievement, not needed before NCLB. To achieve such testing and to prepare children for such assessments, some school districts are adopting web-based instructional management systems to assist teachers, principals, district administrators, and parents in managing and understanding how all the information provides guidance for differentiated instruction in classrooms (Holcomb, 2004).

The complexity of an innovation may negatively impact adoption. The greater the learning curve for using a new software application by a classroom teacher, who must decide between activities that compete for his limited amount of time with students daily and weekly throughout the school year, the less likely the teacher will take up that innovation. The same is true for the trialability of an innovation. If they can not try it out before they buy it, the chances are teachers will move on to the next option that they can try out. Similarly, the more likely the results of the innovation are obvious and observable to others (e.g., ease of development of websites using new web development software) the more likely others may adopt the innovation.

Rogers notes that "[t]he diffusion of innovations takes place within a social system" (Rogers, 1962. p. 303). It is within this social setting that individual users of an innovation/change show others how wonderful or awful the innovation/change is for them setting off a chain of reactions either toward more individuals taking up the innovation/change or rejecting it. He further notes that the status of the individual offering the innovation/change will effect whether others adopt or reject, with the change agent or facilitator of change is an important factor in the process. So, if someone teachers respect and trust is the one providing the evidence, they are more likely to listen than if the individual is someone they distrust. This can be a factor whether a top-down or bottom-up process of change is adopted.

These criteria can be used to assess implementation of any innovation or change, from a new curriculum to the use of computers in classrooms. Rogers' early work in 1962 was supported later by that of Katz, Levin, and Hamilton in 1963. Based on the work of Rogers (1962) and Katz et al. (1963), Guba (1968) noted that the key element of diffusion of an innovation was acceptance by the adopting unit, whether it be a school district, school, or teacher. Guba further suggested that the "seller of an innovation" should use diffusion techniques like telling, showing, helping, involving, and training, all related to the five criteria posited by Rogers (1962). Further, any assumptions about the adopter should be acknowledged, for example, any biases or preconceived notions.

Michael Fullan and Alan Pomfret (1977) specifically looked at the implementation of innovation or change in curriculum and instruction practices. They noted that implementation is not an extension of planning, but is what happens between the design process and actually carrying out that design by various people who probably did not design the innovation. They listed five dimensions of change within implementation of an educational innovation, specifically curricular or instructional: changes in materials, structure, roles and behaviors, knowledge and understanding, and in value internalization. Within each educational innovation, these dimensions would play out at differing levels over time.

Given complexity of the implementation process, Fullan and Pomfret (1977) further explained "determinants of implementation." First, the "characteristics of the innovation" must be considered, including the explicitness of the who/what/when/how questions and the complexity of the innovation. The more complex the project, the more variance there will be in implementation. Secondly, the "strategies for implementation" are important, including what type of training is involved, what resource support is available (time and materials), what feedback mechanisms are in place, and the level of participation. The combination of these four strategies is important, not any one of them independently. Thirdly, the "characteristics of the adopting unit" must be considered. What was the adoption process? Was it through a problem-solving approach or opportunism? What is the organizational climate regarding teacher morale, principal support, and administrative support? What is the environmental support? Fourthly, what are the demographic factors, including school level (elementary or secondary), social class, presence/level of conflict or apathy, level of needs, and competing programs? Clearly, competing agendas will cause problems at many levels. Lastly, the "characteristics of the macro sociopolitical units," or the outside agencies, must be considered. At this level, issues of the design of assessment of the project (e.g., voluntary/mandatory participation, random assignment to test/control groups), incentives for implementation, the role of evaluation (formative or summative), and political complexity of implementing a large-scale project come into play. Each of these determinants will play a role in whether the innovation is successfully implemented or not, and on whether adoption turns into institutionalization.

Paul Berman and Milbrey W. McLaughlin (1978) reported as part of a Rand® Corporation study team looking at federal change agent grants in the 1970s. They noted several factors that affect the fate of project outcomes, which included professional development of teachers about the innovative programs being implemented. "The fate of an innovation depends on the complex interplay among characteristics of the innovative project itself and the institutional setting it seeks to change" (Berman & McLaughlin, 1978, p. 22). Three characteristic phases of the process of innovation were described. The mobilization phase includes the why and how of the project, including the patterns of support for the project (e.g., opportunism, top-down, localized, or broad-based). The implementation phase involves the "translation of the project plans and proposals into practice" (p. 15). They reported that "adoption" occurs with implementation, so no two sites implement alike. Implementation was described as resulting in one of three outcomes: non-implementation where the project basically broke down or retention of status quo with no change; cooptation where the staff adapted the project as they saw fit, not implementing as written; and mutual adaptation where both the project and the setting changed to support outcomes. Mutual adaptation was noted as the only process that led to change in teacher practices. Lastly, the institutionalization phase puts the project as part of the natural practices in the school. Unfortunately, many of the projects they reviewed were never institutionalized. They noted many factors leading to discontinuation or disjointed continuation, including budget constraints and lack of full teacher or administrative buy-in.

One other aspect of the Rand Corporation change agent studies was discovering that many of the projects reviewed had been opportunistic in nature. The schools or districts had not done a systematic review of innovation/change options or a systematic review of current practices to determine the need for change or select the best option to solve a known problem. On the contrary, the studies discovered that projects were undertaken "for reasons fundamentally unrelated to the educational delivery needs of the districts" (Berman & McLaughlin, 1978, p. 14). Strategies that were listed as ineffective in implementation of innovations/change were use of outside consultants, packaged management approaches, one-shot pre-implementation training, pay for training, formal evaluation, and comprehensive projects that were too much and too soon. Effective strategies were the following (p. 34):

- Concrete, teacher-specific and on-going training,
- Classroom assistance from project or district staff,
- Observation of the project in other classrooms or districts,
- Regular project meetings,
- Teacher participation in project decisions,
- Local development of materials, and
- Principal participation in training.

One could argue that these practices would meet the criteria set forth by Rogers (1962, 1972, 1983, 1995, & 2003) for diffusion of an innovation, and those by Fullan and Pomfret (1977). These strategies, along with other findings from the Rand studies will be helpful in analyzing what happened in the Continuous Practice Improvement project and why teachers were continuing to use the project principles three and seven years after participation.

In the book *Taking Charge of Change* (Hord, Rutherford, Huling-Austin, & Hall, 1987), the authors propose that before change can happen in schools, teachers need to pass through several stages of concern. Only when teachers reach the top "impact" stages of concern will the innovation be implemented appropriately. This work refers to a conceptual framework called the Concerns Based Adoption Model (CBAM), first proposed in 1973 by Hall, Wallace, and Dossett. The stages, as numbered by the authors, are (p. 31):

0. **Awareness:** Not aware of the innovation and not concerned about it,
1. **Informational:** Aware of the innovation and wanting more information,
2. **Personal:** Concern about how the innovation will effect "me,"
3. **Management:** Concern with amount of time needed to get ready and manage the innovation,
4. **Consequence:** Concern about how the innovation affects students,
5. **Collaboration:** Concern about coordinating what I'm doing with other teachers, and
6. **Refocusing:** Have ideas to improve the innovation.

The first three stages were described as for the "self," with the teacher focused on her own use. The fourth stage was described as "task" oriented, with the teacher focused on getting the work done. And the last three stages were described as "impact" stages, with the teacher focused on how the innovation has an impact at different levels on students.

To further complicate the conceptual framework called CBAM, Hall and Hord (1987) proposed that there are levels of use that need to be considered as well. While they again propose seven levels of use, they note that the last two may never be reached. The levels of use, as Hall and Hord designated, are (p. 55):

0. **Non-use:** Little to no knowledge of the innovation and doing nothing to learn about the innovation,
1. **Orientation:** Actively getting information about the innovation, including what is required of them and their students,
2. **Preparation:** Actively preparing to use the innovation,
3. **Mechanical use:** Focusing efforts on day-to-day use of the innovation,

4a. **Routine use:** Active, ongoing use of the innovation within classroom routines, less preparation and thought is needed to implement the innovation,

4b. **Refinement:** Varying use of innovation as needed for the students,

5. **Integration:** Integrating the innovation into activities with colleagues for a collective impact on students, and
6. **Renewal:** Effecting changes in the innovation through modifications and alterations.

In reviewing the two lists, one recommendation would be to attend to both sets of issues when introducing an innovation and assessing the impact over time. With every attempt at change in schools, time is always a factor—too little time available to implement the innovation and/or too much time needed to determine full impact.

In the book *Change in schools: Facilitating the process* (Hall & Hord, 1987), the authors argue that the facilitator of change needs to be aware of teacher concerns and point of view. Change is a process, not an event; therefore, the facilitator of change needs to anticipate, as much as possible, what might occur during the change process. This information would be quite powerful for the facilitator to be successful. They suggest that only after the facilitator of change understands how each individual teacher experiences a change process can they plan the change process they are to present. This process takes the philosophy that change is a social process, not just an individual process (not antithetical to that of Rogers). "It is the dynamics of the group, not the skills of its individual members, that is both the major source of problems and the primary determiner of the quality of solutions" (Schmuck, Runkel, Arends, & Arends, 1977, p. 3). However, successful implementation of an innovation/change goes way beyond the concerns of the teacher to the motivation for the change as noted in the Rand Change Agent studies of the 1970s.

In 1982, Fullan, in the first edition of his book on change in schools, put forth 10 factors affecting adoption of an educational change. These were:

1. The existence and quality of the innovation
2. Access to information
3. Advocacy from central administration
4. Teacher pressure/support
5. Consultants and change agents
6. Community pressure/support/apathy/opposition
7. Availability of federal or other funds
8. New central legislation or policy (federal/state/provincial)
9. Problem-solving incentives for adoption
10. Bureaucratic incentives for adoption

This list is very clear and straight forward. While the factors are not listed in any particular order, each has been demonstrated in the literature to have an impact on adoption and the level of adoption that might occur.

In the 2001 edition of his book, Fullan posits three interactive factors that affect implementation of an innovation or educational change. These are:

1. The characteristics of the change (e.g., the need for the change, the clarity of the goals and means of the change, the complexity and extent of the change required, and the quality and practicality of the change);

2. Local characteristics (e.g., the district, the community, the principal, and the teachers); and
3. External factors (e.g., government and other agencies).

While addressing each of the factors may seem obvious in planning for change, Fullan points out in each instance there are components of the process that complicate the process of implementing change. And these also address the issues affecting continuation of the innovation after it has been implemented and initially assessed. "Lack of interest or inability to fund 'special projects' out of district funds and lack of money for professional development and staff support for both continuing and new teachers signaled the end of many implemented programs" (Fullan, 2001, p. 88). Often an innovation is put into place because there is external funding available (e.g., Preparing Tomorrows Teachers to use Technology, or PT3, grants, and Technology Literacy Challenge Grants); however, when the funding moves to another emphasis (e.g., from early literacy to middle grades literacy or math), the initial innovation may go by the wayside as a victim of the seemingly constant changing of the funding agency's priorities. He also notes that giving an innovation the "special projects" category often dooms the project to extinction even if it works fabulously. "Special projects" are targets for budget cuts since they are considered non-essential. It is very hard to defend continuation of a "special project" as compared to essential projects.

Summary

There is no simple or single-factor theory for change. Change in general, as well as in schools, has multiple layers of complexity. Each of these approaches to the study of change, successful change or not, have some common threads—having knowledge about the innovation, wanting to learn more about it, getting ready to give it a try, trying it out with your own students, and possibly adapting it to the needs of your students. This process fits with the "mutual adaptation" concept put forth by Berman and McLaughlin (1978) where both the innovation and the site of the innovation change in mutually supportive ways. Being able to communicate the significance of the innovation/change to the relevant stakeholders (e.g., principals, teachers, parents, students), so it is clear that the innovation/change is not the enemy, is essential. For an innovation/change to be successful, all affected parties must work together to support each other through the process of giving up one practice and taking up a new one. Change is difficult and painful, but can be managed successfully to support all parties concerned.

References

Berman, P., & McLaughlin, M.W. (1978). *Federal programs supporting educational change, Volume VIII: Implementing and sustaining innovations*. Santa Monica, CA: Rand.

Cuban, L. (1986). *Teachers and machines: The classroom use of technology since 1920*. New York: Teachers College Press.

Fullan, M. (1982). *The meaning of educational change*. New York: Teachers College Press.

Fullan, M. (2001). *The new meaning of educational change.* Third Edition. New York: Teachers College Press.

Fullan, M., & Pomfret, A. (1977). Research on curriculum and instruction implementation. *Review of Educational Research, 47*(1), 335-397.

Guba, E.G. (1968). Diffusion of innovations. *Educational Leadership, 25*(4), 292-295.

Hall, G.E., & Hord, S.M. (1987). *Change in schools: Facilitating the process*. Albany, NY: State University of New York Press.

Hall, G.E., Wallace, R.C., & Dossett. W.A. (1973). *A developmental conceptualization of the adoption process within educational institutions. (Report No. 3006).* Austin: The University of Texas at Austin, Research and Development Center for Teacher Education. (ERIC Document Reproduction Service No. ED095 126).

Holcomb, E. (2004). *Getting excited about data.* Second Edition. Thousand Oaks, CA: Corwin Press.

Hord, S.M., Rutherford, W.L., Huling-Austin, L., & Hall, G.E. (1987). *Taking charge of change.* Alexandria, VA: Association for Supervision and Curriculum Development.

Katz, E., Levin, M.L., & Hamilton, H. (1963). Traditions of research on the diffusion of innovations. *American Sociological Review, 28*(2), 137-152.

Little, J.W. (1993). Teachers' professional development in a climate of education reform. *Education Evaluation and Policy Analysis, 15*(2), 129-151.

Papert, S. (1993). *The children's machine: Rethinking school in the age of the computer*. New York: Basic Books.

Rogers, E.M. (1962, 1972, 1983, 1995, 2003). *Diffusion of innovations,* Fourth edition. New York: The Free Press.

Schmuck, R.A., Runkel, P.J., Arends, J.H., & Arends, R.I. (1977). *The second handbook of organizational development in schools.* Eugene, OR: Center for Educational Policy and Management.

U.S. Department of Education. (2001). *No Child Left Behind Act*. Washington, D.C.

Chapter III

"Making Teachers Better": A Brief History of Professional Development for Teachers

One of the requirements of a profession is that its members somehow continue to learn, to grow, to renew themselves, so that their interactions with ideas and with clients are reflective of the best knowledge and skill available to them (Griffin, 1979, p. 127).

In 2007, reports from the Office of the Secretary of Education's Web site listed statistics about educational technology and schools. As of November 2006, the Internet Access in U.S. Public Schools and Classrooms: 1994-2005 report indicated "the ratio of students to instructional computers with Internet access in public schools was 3.8 to 1..." This was a decrease from 2003, when the ratio was 4.4 to 1 (Department of Education, 2006). There was also information about funding, including the $2.25 billion in the federal E-rate program supporting discounts on telecommunications services, Internet access, and networking for schools and libraries. With each federal funding program for educational technology, professional development was an essential component, requiring up to 25 percent of funds to be used for professional development on using technology to improve student outcomes.

Professional development for teachers has been around for decades with mixed results. One major reason for professional development for teachers has been to effect change in practices in classrooms, either through policy changes or curriculum changes, among other forms of reform.

This chapter will look at the history of general professional development for teachers over the last 30 years. There is a rich literature on this topic, so this chapter is not exhaustive, but attempts to provide highlights into recent history. Given the fact that microcomputing is but 30-plus years old, and microcomputers have not been in most schools for that long, the relatively short history of professional development for teachers in using computers and infusing computers in classroom teaching activities will be discussed.

Professional Development for Teachers

Through the 1960s and 1970s, professional development was primarily delivered through inservice teacher training, which consisted of a series of "[d]isjointed workshops and courses focus[ed] on information dissemination rather than stressing the use of information or appropriate practice in the classroom" (Wood and Thompson, 1980). Wilson and Berne (1999) described these events as scattered and serendipitous, rather than planned and deliberate. These half-day or full-day events continued to be common in public education, in spite of efforts of staff developers to persuade school district officials to use other methods.

One method of providing professional growth for teachers has been through extended workshops, defined as a series of after school, two-hour, weekend-long, or several day-long workshops, sometimes for one or two weeks during the summer months. During these sessions the teachers "learned" new methods of teaching particular subjects, or a new philosophy of teaching a particular subject (e.g., early balanced literacy versus phonics based teaching). In both of these instances, inservice training or extended workshops, the process of helping the teachers learn new methods of teaching was typically taken out of context, even when the training was within their own school. Often these sessions presented new methods with theory to be learned (e.g., constructivist teaching and learning), a presentation of methods through videotape, film, or a slideshow, and perhaps some hands-on experiences as a whole group of teachers. But, the discussion was not about how this new method, theory, or policy was useful to the third grade teacher of mentally gifted children specifically in his classroom or the fifth grade teacher of special education children in her classroom. In other words, the training was generic and it was left up to the teachers to make use of the information or skills in his/her classrooms.

Since 1980 there have been any number of innovative practices brought to bear on teachers' classrooms, from the inclusive classroom to the standards movement to

Howard Gardner's learning styles to incorporation of computers into teaching and learning, to name but a few. But has professional development for teachers really changed much since the 1980s? The research and practice literature would suggest there has been little change. Smylie (1989) reviewed a national survey of teachers regarding their views on the effectiveness of different sources of learning how to teach differently. From a sample of 1,789 teachers, direct experience in the classroom was considered the most effective source of learning how to teach differently. The least effective was inservice teacher training. Of most importance to these teachers was for the knowledge they were learning to be relevant to their classroom context and overall practice.

A number of authors have written articles giving advice to school districts on how to make professional development successful. Sparks (1994) discussed a paradigm shift in staff development. He suggested three factors for changing teacher professional development from the old traditional model to the new. He posited these sessions be about results driven education, noting that student outcomes and not courses taken or grades earned should be the focus. He suggested professional development should be about systems thinking, looking at the sum of the parts of the professional development program, not at the individual pieces. And lastly, Sparks suggested teacher professional development should be about constructivism, with collaboration with peers, action research back in classrooms, and overall reflective practices regarding what worked, what did not work, and why. All of these suggestions are like motherhood and apple pie in the era of the No Child Left Behind Act of 2001. But achieving these lofty goals may not be as easy as they sound given the differences between school districts and schools still being described as "loosely coupled" (Sun-Keung Pang, 2003; Weick, 1976).

Abdal-Haqq (1995) provided a list of 11 characteristics for effective professional development. Lauro (1995) compared five approaches to professional development, but stressed that, depending on the program, school, or district, complete implementation could take up to five years. Guskey and Sparks (1996) suggested that successful staff development happens when the magnitude of change is not so large as to require coping strategies that may compromise the changes, but large enough to require real effort.

Another concept discussed in the literature about professional development for teachers is how teachers learn. How can you develop a professional development program without understanding how teachers learn? How teachers learn goes to the very essence of who teachers are and how they practice. Ball (1996) listed nine factors considered to impact teacher learning. These were:

1. Prior beliefs and experiences about the process of learning that shape future learning, bringing the teacher's own professional history into play.

2. Subject-matter knowledge level, where the teacher's own command of, and confidence in using, the subject-matter is critical for supporting new learning.
3. Knowing the students, their world view and perspectives on learning.
4. Understanding the context (inner city, rural, private or public education), not just how these factors inhibit efforts, but how they facilitate as well.
5. Competing demands on time, when new ideas or practices require unlearning old ideas and ways of practice and then learning the new.
6. Ability to reflect on what works and what doesn't and what facilitates the working and what inhibits.
7. Follow-up on training, through long-term support, coaching, and/or ongoing interaction with peers.
8. Modeling of behaviors; however, Ball reported this as "vague and variously interpreted" (Ball, 1996, p. 502)
9. Teacher control of the agenda, determining the program scope and sequence.

Basically, Ball was suggesting a whole-system approach to professional development, not conducting training out of context or in a vacuum. This process is congruent with what Hall and Hord (1987) proposed for the change facilitator to know and understand the teachers and how the teachers perceived and negotiated change in order to plan accordingly and be more effective. Systemically, if teachers are already insecure about their knowledge of a subject (e.g., mathematics for elementary teachers), professional development about a new curriculum in that subject will be difficult and time consuming. Similarly, knowing the students, the context of the school, and the competing demands on time will allow the change facilitator to be prepared for these challenges. And modeling behaviors through mentors, allowing the teachers to control the schedule, scope and sequence, will be important over time.

Wilson and Berne (1999) offered three themes in contemporary professional development and research on teacher learning. First, communities of learners were redefining teaching practice. No longer should the teacher be alone in the classroom without collegial support. Second, teacher learning should not be bound and delivered, but activated for maximum effect. There should be a movement away from dissemination of information to actively engaging teachers with the information and creating knowledge that will impact the classroom. (Interestingly enough, this approach is in concert with the philosophy of constructivist learning for students, creating an active process for learning in relevant contexts.) Lastly, support of honest and critical collegiality and discussion between teachers was essential to shaping the community of learners. These three themes come full circle. Wilson and Berne

(1999) went further to note that outcomes of professional development depend on many challenges, which resonated with those put forth by Little (1993).

Putnam and Borko (2000) discussed the relationship of "situated perspective" on learning and knowledge. Using the work of Lave and Wenger (1991) and other situated learning theorists, they argued that knowing and learning are situated in both physical and social contexts—social in nature and distributed across persons and tools (p. 12). They tried to recast the relationship between what people know and the setting in which they know, which relates back to Ball's list of factors that impact teacher learning (1996). Professional knowledge is not developed in a vacuum, but in a context relevant to the knowledge, organized and accessed in meaningful ways as relating to the classroom for teachers.

Guskey (2000) points out that professional development has never been as important as it is now for teachers. While his book was published before the No Child Left Behind Act (U.S. Department of Education, 2001), his prediction is most poignant. Teachers are being held accountable for every student learning at proficiency levels in all grade levels—a Herculean task, if ever there were one. States are requiring teachers to accumulate a number of professional development hours or credits each year, or over a number of years, to maintain their teaching certification. Unfortunately, Guskey points out that the state requirements translate into "How can I get my hours?" rather than "What programming options will help me be a better teacher in my classroom" (p. 15). This exemplifies, however, the deficiency approach, saying teachers need to improve since they are not good enough.

Guskey (2000) offers three defining characteristics of professional development, as intentional, ongoing, and systemic. Professional development as an intentional process means that there is clear direction, purpose and goals for the program, that the goals are worthwhile to all participants, and that the evidence of outcomes is known up front. Professional development as an ongoing process assumes that change occurs over time, reinforcing previous activities, reflecting on the relationship to classroom activities, and providing opportunities to collaborate with peers. And, professional development as a systemic process ties all the organizational variables together to support the teachers in their enterprise. When viewed systemically, professional development is seen as improving the entire organizational capacity. Lastly, he points out the major models of professional development as training, observation/assessment, involvement in a development/improvement process, study groups, inquiry/action research, individually guided activities, and mentoring. He notes that each model has "different assumptions, expectations, and beliefs about professional growth," (p. 28) and that because of the differences no single model will be effective for all teachers.

In 2001, the National Staff Development Council generated standards for staff development, which addressed four aspects that should be considered in developing any professional development program. The context addresses the culture of the school, district, and administration leadership. The process is the design of the program. The content addresses the topics considered to be appropriate for the staff development program. Lastly, the most important aspect, and most difficult to ensure, is time for the organizational change process, which can take several months to several years depending on the scope of change being sought. As noted above, Lauro (1995) indicated it could take up to five years for an innovation to be fully implemented and institutionalized. In this vein, Joyce and Showers (2002) discussed the need in teacher staff development to establish a community of teachers working together in support of change in the institution, curriculum, and teaching methods—all with the goal of improving student achievement. They describe a community of teachers as one that "[e]ngages, for a dozen days during the school year, in the formal study of a curriculum area or a teaching strategy that is useful across curriculum areas, and regularly studies implementation and consequent student leaning, then the odds are that student achievement will rise substantially" (Joyce & Showers, 2002, p. 3).

Four conditions need to be met for this to occur. First, there needs to be "a community of professionals who study together, put into practice what they are learning, and share the results" (Joyce & Showers, 2002, p. 4). Within this work, the content is based on the curriculum and related instructional strategies to have a high impact on student achievement. In the end, the processes of staff development empower the teachers to implement what they are learning and to share these experiences with colleagues. The outcomes should be new attitudes, new knowledge and skills that can be transferred directly to the classroom, and new strategies for teaching. The notion of a community of teachers who collaborate in this fashion may not be unique, but it is a powerful concept given the goal of enhancing student achievement. Within the model of a community of teachers fits the concept of peer coaching. Peer coaching is not a new concept, but is supported in the literature by Berman and McLaughlin (1978), Fullan and Pomfret (1977), Sharan and Hertz-Lazarowitz (1980), and Joyce and Showers (1980, 1982).

In addition, Joyce and Showers suggest that between 1980 and 2000 two notable changes occurred in teacher staff development efforts. In their work with schools and districts, they noticed an increase in the duration and intensity of many training programs, including forms of continuing support and follow-up activities. Secondly, the one-shot events are more likely to be noted as "for awareness only" (2002, p.69).

More recently, DuFour (2004) discussed best practices for professional development for teachers. He again noted that historically staff development has been an occasional event, and often away from the classroom or school. He suggested that staff development should stress an increase in a staff's collective capacity, not the individual; challenge staff to act in new ways, building on shared knowledge to apply new concepts in classrooms; focus on results, not activities (e.g., student learning outcomes); and demonstrate sustained commitment to achieving goals.

Lastly, DuFour noted that staff development of teachers must include patience with persistently sustained effort over time. What is noteworthy is how many of these suggestions resonate with all the literature on change. What is also noteworthy is the lack of discussion of change theory in discussion of professional development of teachers, given that all professional development is about affecting change in teacher attitudes, beliefs, behaviors, and practices.

In summary, the question raised earlier was "has professional development for teachers really changed since the 1980s?" From the articles reviewed and the advice given therein, the answer is "perhaps." According to Joyce and Showers, there have been changes in the process, including prefacing the short, one-shot sessions as informational rather than an end in themselves. Movements to include teachers in the planning to make any staff development more relevant to classroom needs will, more likely, affect some level of change in practice. Because change is not welcome, the more teachers are part of the process of planning the programs, the more likely they will be supportive of the change process. As Guskey (2000) pointed out, and with the No Child Left Behind Act of 2001, teachers are being held accountable for student learning, so it is imperative that staff development be directed toward effecting student achievement, as described by Joyce and Showers (2002).

Professional Development and Computers

Looking back at professional development and computers, there really was nothing in place in schools or districts in the early years of microcomputers. What was available came from the vendors, such as Tandy Corporation for use of the TRS80 computer, or from other users through user groups for programming in LOGO. Otherwise, those interested in computing in the early 1980s were basically left with either attending computer courses in a university that were geared toward programming in languages and not teaching with microcomputers, or learning on their own.

Having said that, teachers began to learn about teaching with computers in classrooms when Apple, Atari and Commodore computers began to come into classrooms in the late 1970s. Early uses of computers were for drill and practice of skills, and for learning programming languages like LOGO. For many teachers, computer training was not about how to use the computer to teach, but how to use the computer to support teaching through CAI, an approach that was a hold-over from mainframe days of computing in schools described in Chapter I. Overall, the computer was relegated to an add-on to the classroom by teachers who were not the "early adopters." Computers often were limited to a computer lab where a teacher (maybe a computer teacher and maybe a teacher who knew more about computers than anyone else in the school) taught computer programming (BASIC or LOGO) or delivered programmed instruction. Hence, classroom teachers were not required to learn about using computers with their own students.

The Office of Technology Assessment (OTA), in its 1988 report *Power On!,* noted that only one-third of all K-12 teachers had as much as ten hours of computer training, and "much of this had been focused on learning *about* computers, not learning how to teach *with* computers" (p. 18, emphasis original). By this date, the OTA report noted that all teacher education programs offer some level of computer training, but resources limit experiences beyond introductory courses. Along with this information, teaching with computers was seen as making teaching more challenging rather than facilitating teaching.

In 1985/1986, Apple Computer Company partnered with eight school districts around the United States and launched the Apple Classrooms of Tomorrow (ACOT^{SM}) project. The ACOT project put computers into the hands of teachers and students in the regular classroom. Initially, there were two computers for each student and teacher, one at his/her desk at school and one at home. The ACOT studies lasted over ten years and provided significant insight into how teachers learn to infuse computers into classroom teaching and learning activities. One specific outcome from the studies was that the more things change (in teaching and classrooms) the more teachers must confront previous beliefs about learning. Change in instruction only happens with a corresponding change in beliefs about instruction and learning, which means taking risks and the possibility of failure (Dwyer, Ringstaff, & Sandholtz, 1991). The ACOT studies showed time and again how teachers who had regular access to computers in their classrooms over several years, with sustained support throughout the experience, changed their instruction significantly. Dwyer et al. (1991) reported that the teachers who came into the project "believed that technology would make their jobs easier and more efficient. Most never dreamed they would alter their instructional approaches" (p. 50). These teachers were reported to have moved away from curriculum-centered instruction toward child-centered instruction, and toward active rather than passive learning.

Over the years of the ACOT studies, a pattern of teaching and learning emerged at all sites demonstrating that teachers go through five stages of change in practice to fully integrate computers into classrooms. The Entry Stage signified teachers with little or no experience with computers. Teachers were learning how to actually use a computer, and used the computer to support the familiar world of teaching. The Adoption Stage demonstrated teachers were no longer afraid of computers and were willing to use it to support text-based and drill-and-practice instruction. In the Adaptation Stage, the new technology was used throughout the curriculum. Student productivity and the quality of student engagement in the classroom appeared to increase. In the Appropriation Stage, the teachers were so comfortable with their own mastery of the computer, they were ready and able to give the students license to use the technology in new and exciting ways, such as through project-based learning activities. The final stage was Invention where clearly the teachers and students were using computers in ways they had never considered possible, as in creating Web pages in the latter years of the ACOT project.

While the ACOT studies demonstrated how teachers can develop from non-users of computers to experienced innovators, the overall program must be highlighted. Not only did each participating student and teacher receive two computers, one for home and one for their classroom desktop, but the teachers also gained access to videodisks and videodisk players, video cameras, scanners, CD-ROMs, and modems to access the Internet and online communication. Students were using an integrated system that provided word processing, database, spreadsheet, and graphing capabilities. The support provided by the ACOT staff was tremendous and constant.

Clearly, the experience of these teachers is not the typical experience of classroom teachers learning how to use computers with their students in their curriculum. As Yocam and Wilmore (1994) indicated, the ACOT project took the stance that staff development needs to be grounded in context of practice, to see expert teachers modeling instructional practices as they work with students, and to provide time to reflect on observations and to discuss instructional and learning principles at work. This project also demonstrated that teachers with extensive support from and access to experts in using computers in the classroom are more likely to achieve a high degree of self confidence and use of computers with their students. As stated above, the teachers who came into the project "believed that technology would make their jobs easier and more efficient" (Dwyer et al., 1991, p. 50). Participants were not randomly assigned to research and control groups to see what variations on the support provided might have achieved for both the teachers and the students. But then, most professional development programs are not based on research design.

David Hurst, in 1994, noted that "If teachers are to use technology effectively in their classrooms, we must give them adequate inservice training programs." While his declaration seemed novel at the time, he was merely acknowledging the lack of support teachers had received in the eighteen years since microcomputers entered classrooms. In 1994, the generation growing up would be weaned on video games, VCRs and home computers, thus expecting classrooms to reflect the rest of the technology rich world. By the time of his article in 1994, Hurst was suggesting that teachers needed to be proficient in word processing, databases, spreadsheets, desktop publishing, electronic communication, and multimedia. (p. 74) By the early twenty-first century, teachers not only needed to be proficient in these technologies, but also in how these tools and digital audio and video support their teaching and their students learning to result in deep understanding and knowledge – going beyond knowledge *of* the tool to *how* the tool to supports learning.

As with general professional development sessions for teachers, sessions about technology have been approached in a similar fashion. With school districts investing great sums of resources to put computers and Internet access into schools, it is essential teachers know how to take advantage of these resources. Hurst (1994) noted that "[a]n intensive, one-day inservice the day before students arrive for full classes is not the best approach, and yet I see this happening year after year in school systems across the nation" (p. 75). He was arguing for ongoing training for teachers to become comfortable with using technology before asking them to be comfortable using it with students. Somehow the concept of ongoing training sounds familiar. Similarly, Schofield (1995) noted that there were a number of problems with professional development around using computers in classrooms, specifically inappropriate timing, a lack of coordination between training and hardware, a lack of matching training with teacher knowledge and instructional concerns, and a lack of concentrated experiential learning opportunities. Answering these concerns supported Hurst's concerns for ongoing training that was focused on teacher needs.

Grant (1996) provided an in depth review of professional development for technology as compared to the "training paradigm" traditionally used. She suggested that effective professional development programs were linked to building a professional culture in schools, including having a vision of technology as an empowering tool for teachers and students. Unfortunately, many technology training programs focused on how to make the technology work, not on how the technology benefits children's learning. From her review of the literature and an Office of Technology Assessment report (1995), entitled *Teachers and Technology: Making the Connection*, Grant listed five supports needed for teachers to use technology effectively in classrooms:

1. Access to equipment, in good condition and working
2. Technical and pedagogical assistance, ongoing, on-site and job embedded
3. Administrative and community backing and commitment
4. Time for learning and collaboration
5. Sustainability through systemic reform

One can clearly argue that all of these supports were in place for the ACOT teachers as described above. However, we also noted that what was afforded the ACOT teachers was not the norm in the majority of schools. Teachers are not, for the most part, technologically advanced so for them having computers in good condition and working is the bare minimum. At the same time, when there is a problem, having ongoing, local support is critical. Teachers do not want to be prepared to teach a lesson with a computer as the primary tool only to learn that the computer has stopped functioning or the server is down. These experiences will often cause teachers to shy away from using computers at all in lessons. In order for these two conditions to be met, administrative and community support are needed. Principal endorsement of computing in classrooms and the curriculum is key to teachers seeing and making the connection. Backing from the principal needs to include also providing time for teachers to learn independently and collaboratively with others in the school. These support systems will provide the basis for sustainability because they are systemic in nature.

Brand (1997) added a number of additional factors that are needed to ensure success in professional development on educational technology. Beyond the need for time to learn and transfer skills into classrooms, planners need to take into consideration the varying requirements of the teachers and develop flexible programs given different skill levels and needs. In addition to ongoing technical and pedagogical support, planning should provide a collaborative, cooperative, non-threatening environment for learning, and make the link between technology and educational objectives. These concepts were supported by Meltzer and Sherman (1997); Peterman, McGilliway, and Franz (1998); Eli (1999); Cooley (2000); and Bitner and Bitner (2002).

As noted in the review of literature about general teacher professional development, the use of technology in teaching and learning was all about use in the classroom. Maddin (1997) suggested learning from teachers as they are teaching in their classrooms, where novice technology-user teachers could observe experienced technology-user teachers in practice using learner-centered pedagogy. He borrowed the term "shadowing" from business, and suggested hiring substitutes for classrooms to allow the novice teacher to learn from the experienced teacher.

Kleiman and Johnson (1998) suggested moving toward looking at student outcomes rather than focusing on individual teacher skill development. Moving from the research literature to the classroom, they suggested professional development around technology should consider the teachers' beliefs and preferred practices of teaching; what type and how much technology is available in the classroom; and the level of students and content being taught. Lastly, Kleiman and Johnson (1998) suggested, from information about a capacity building project in Boston public schools, that teachers want to see what other teachers are doing in order to consider how to change their own practice.

Sandholtz (2001) compared two teacher professional development programs related to teaching with technology. One was the ACOT Teacher Development Centers and the other was a program developed by a southern California school district. She noted that frequently funding goes toward equipment rather than training the teachers in using the technology in the classroom. Also, she suggested that student teachers, who learned the latest technology in technology education, "are unlikely to integrate technology into their own teaching if they don't observe effective technology practices in schools" (p. 351). From her comparison of the two programs, she noted eight common components for effective professional development around technology. These were:

1. Teacher input into design, demonstrating value of the teacher perspectives and that teachers are key to determining what they need to change practice.
2. Teacher choice to participate and when as appropriate to her needs.
3. Administrator involvement with the principal actively involved in planning how to support the teachers as they begin to implement new practices in the classroom.

4. "Situated teacher development" allowing teachers to see expert teachers modeling practice as they work with children.
5. Participant collaboration through teams, teachers teaching teachers, and group reflection.
6. Constructivist environment through hands-on exploration and practice, all in an non-threatening environment.
7. Flexibility of training to meet the needs of teachers.
8. Adequate funding, since all of this is quite expensive.

While all of these components seem obvious from reviewing the literature, they are not easy to achieve. Principals who are not yet convinced that technology training is important for teaching and learning will be hard pressed to be actively involved. Situated learning, a la Lave and Wenger (1991), demonstrates the importance of seeing learning through the lens of the situation and how observing masters at work facilitate novice learners. Constructivist approaches to teaching and learning do not have to be at odds with the No Child Left Behind Act's approach to testing and standards accountability, but the law does have the potential to drive instruction toward a teacher-oriented approach to teaching and learning, which *is* at odds with constructivist theory.

Zhao and Cziko (2001) presented the Perceptual Control Theory as another way of considering developing professional development programs for teachers (p.6) They suggested three conditions for teachers to use technology: (1) the need for teachers to believe that technology can more effectively meet a higher-level goal than what they have been using, (2) that using technology will not cause disturbances to other higher-order goals, and (3) that he or she has or will have sufficient ability and resources to use the technology. They suggest that districts have made the assumption in the past that, if the teachers know how to use the technology, they will use it in teaching, which simply has been proven wrong. If one adheres to the three conditions listed by Zhao and Cziko, greater success would be attained from professional development programs around technology.

Bradshaw (2002) drew upon the work of Joyce and Showers (1981) about staff development programs for teachers. He suggested a comprehensive staff development program for technology in teaching and learning should consider the four "incremental" categories of staff development activities proposed by Joyce and Showers. These were (1) present theory, (2) present theory and model through a demonstration, (3) present theory, modeling and opportunities to practice in a non-threatening environment, and (4) present theory, modeling, opportunities to practice, and follow-up through coaching, study groups or peer visits. Next he suggested using the seven stages of use of the Concerns-Based Adoption Model (CBAM) proposed by Hall and Hord (1987), as described in Chapter II. And lastly, Bradshaw suggested evaluating the outcomes of the program for satisfaction, knowledge and/or skills, classroom use, and student performance. Many professional development programs start and stop evaluation with the workshop satisfaction survey, but do not follow up with an assessment of change in practice.

Mills and Tincher (2003), in developing a technology-oriented professional development program, used the Innovation Configuration Matrix or Map (ICM) from the CBAM model by Hall and Hord (1987). They developed a two-dimensional matrix of the components related to technology integration, as they defined it, from novice user to expert user with change over time and across contexts. Their study demonstrated that technology professional development needed to focus on instructional strategies and methods to integrate technology into lessons rather than on the technology as a tool. They noted that as teachers moved through the stages of integration they realized that technology was more than a mere tool, but something to creatively augment student learning. They demonstrated, like the ACOT studies, that technology integration is a developmental process, and that professional development must be the means for guiding teachers through the stages from novice user to expert by providing appropriate training as needed for as long as needed, noting that technology professional development for most teachers is throughout their professional career.

As Angel noted (2004), technology is accepted as part of life in the twenty-first century, but not in schools. Training still tends to be about "how-to-use" hardware and software, rather than how to teach in a technology rich environment. Teachers tend to be concerned that technology will fail at an important time, thus wasting student learning time. The 11-step staff development model he presented included involving the teachers in the process, gathering hard data before planning, securing support staff, planning the program with time for teachers to "play" with the technology and time to share products, documenting classroom use, and using the information gathered throughout the program for modifications. Oddly enough, his approach resonates with what has been presented over the years, which begs the question of who reads these papers? (One does wonder!)

Summary

From this review of some of the literature on general professional staff development and professional development regarding the use of computing in classrooms for teachers, history seems to repeat itself time and again, or at least developers of technology-related professional development do not seem to learn from history or the literature. If it were otherwise, why would authors continuously "recommend" sustained professional development that is rooted in classroom needs and practice? While authors have routinely noted that one-time, short in duration staff development programs with no follow-up tend to result in little to no change in classroom practice, these types of programs still exist in some settings, especially around technology. As researchers document that the more training in actual infusion of technology options or strategies in classrooms translates into use and student achievement, there are still articles that advise planning for continuity of training over time. While looking for change in the quickest, cheapest, and easiest way possible, programs are developed often to meet specific and urgent needs, rather than looking at the long term impact across all programs, or through a systematic approach to professional development overall. The next chapter will present a professional development model for learning how to infuse computers into classroom practice that addresses most of these suggestions—Continuous Practice Improvement.

References

Abdal-Haqq, I. (1995). *Making time for teacher professional development. (Digest 95-4)*. Washington, DC: ERIC Clearinghouse of Teaching and Teacher Education.

Angel, R. B. (2004, March). A procedural model for planning and delivering technology staff development. Paper presented at the annual SITE Conference in Atlanta, GA.

Ball, D.L. (1996). Teacher learning and the mathematics reforms: What we think we know and what we need to learn. *Phi Delta Kappan, 77*(7), 500-508.

Berman, P., & McLaughlin, M. W. (1978). *Federal programs supporting educational change, Volume VIII: Implementing and sustaining innovations*. Santa Monica, CA: Rand.

Bitner, N., & Bitner, J. (2002). Integrating technology into the classroom: Eight keys to success. *Journal of Technology and Teacher Education, 10*(1), 95-100.

Bradshaw, L. K. (2002). Technology for teaching and learning: Strategies for staff development and follow-up support. *Journal of Technology and Teacher Education, 10*(1), 131-150.

Brand, G. A. (1997). What research says: Training teachers for using technology. *Journal of Staff Development, 19*(1). Retrieved on October 25, 2004 from http://www.nsdc.org/library/publications/jsd/brand191.cfm.

Cooley, V. E. (2001). Implementing technology using the teachers as trainers staff development model. *Journal of Technology and Teacher Education, 9*(2), 269-284.

DuFour, R. (2004). Leading edge: The best staff development is in the workplace, not in a workshop. *Journal of Staff Development, 25*(2). Retrieved on September 10, 2004 from http://www.nsdc.org/library/publications/jsd/dufour252.cfm

Dwyer, D. C., Ringstaff, C., & Sandholtz, J. H. (1991). Changes in teacher's beliefs and practices in technology-rich classrooms. *Educational Leadership, 48*(8), 45-52.

Eli, D. P. (1999). Conditions that facilitate implementation of educational technology innovations. *Educational Technology, 39*(6), 23-27.

Fullan, M., & Pomfret, A. (1977). Research on curriculum and instruction implementation. *Review of Educational Research, 47*(1), 335-397.

Grant, C. M. (1996). Professional development in a technological age: New definitions, old challenges, new resources. Retrieved on July 21, 2004 from http://ra.terc.edu/publications/terc_pubs/tech-infusion/prof_dev/prof_dev-intro.html

Griffin, G. A. (1979). Guidelines for the evaluation of staff development programs. In A. Lieberman and L. Miller (Eds). *Staff development: New demands, new realities, a new perspective*. New York: Teachers College Press, pp 126-143.

Guskey, T. R. (2000). *Evaluating professional development*. Thousand Oaks, CA: Corwin Press.

Guskey, T. R., & Sparks, D. (1996). Exploring the relationship between staff development and improvements in student learning. *Journal of Staff Development, 17*(4). Retrieved on October 25, 2004 from http://www.nsdc.org/library/publications/jsd/f_gusky.cfm

Hall, G. E., & Hord, S. M. (1987). *Change in schools: Facilitating the process*. Albany, NY: State University of New York Press.

Hurst, D. S. (1994). Teaching technology to teachers. *Educational Leadership, 51*(7), 74-76.

Joyce, B., & Showers, B. (1980). Improving inservice training: The messages of research. *Educational Leadership, 37*(5), 379-385.

Joyce, B., & Showers, B. (1981, April). *Teacher training research: Working hypotheses for program design and directions for future study*. Paper presented at the annual American Educational Research Association conference, Los Angeles.

Joyce, B., & Showers, B. (1982). The coaching of teaching. *Educational Leadership, 40*(1), 4-16.

Joyce, B., & Showers, B. (2002). *Student achievement through staff development.* Third edition. Alexandria, VA: ASCD.

Kleiman, G., & Johnson, K. (1998, Sept.-Oct.). Professional development: From reports to reality. *LNT Perspectives, 5*. Retrieved on July 12, 2004 from http://www2.edc.org/LNT/news/Issue5/feature.htm

Lauro, D. R. (1995). Five approaches to professional development compared. *THE Journal Online*. Retrieved on October 25, 2004 from http://www.thejournal.com/magazine/vault/articleprintversion.cfm?aid=1070

Lave, J., & Wenger, E. (1991). *Situated learning: Legitimate peripheral participation*. Cambridge: Cambridge University Press.

Little, J. W. (1993). Teachers' professional development in a climate of education reform. *Education Evaluation and Policy Analysis, 15*(2), 129-151.

Maddin, E. A. (1997). The real learning begins back in the classroom: On-the-job training and support for teachers using technology. *Educational Technology, 37*(5), 56-59.

Meltzer, J., & Sherman, T. M. (1997). Ten commandments for successful technology implementation and staff development. *NASSP Bulletin, 81*(585), 23-32.

Mills, S. C., & Tincher, R. C. (2003). Be the technology: A developmental model for evaluating technology integration. *Journal of Research on Technology in Education, 35*(3), 382-401.

National Staff Development Council. (2001). *Revised standards for staff development*. Oxford, OH.

Office of Technology Assessment, U.S. Congress. (1988). *Power on! New tools for teaching and learning.* Washington, D.C.

Office of Technology Assessment, U.S. Congress. (1995). *Teachers and technology: Making the connection*. OTA-HER-616. Washington, DC: U.S. Government Printing Office. Retrieved on October 28, 2004 from http://www.wws.princeton.edu/cgi-bin/byteserv.prl/~ota/disk1/1995/9541/954104.pdf

Peterman, L., McGillivray, K., & Franz, J. (1998, Nov.-Dec.). Professional development: From reports to reality, Part 2. *LNT Perspectives, 6*. Retrieved on July 12, 2004 from http://www2.edc.org/LNT/news/Issue6/feature.htm

Putnam, R. T., & Borko, H. (2000). What do new views of knowledge and thinking have to say about research on teacher learning? *Educational Researcher, 29*(1), 4-15.

Sandholtz, J. H. (2001). Learning to teach with technology: A comparison of teacher development programs. *Journal of Technology and Teacher Education, 9*(3), 349-374.

Schofield, J. W. (1995). *Computers and classroom culture.* New York: Cambridge University Press.

Sharan, S., & Hertz-Lazarowitz, R. (1980). A group investigation method of cooperative learning in the classroom. In S. Sharon, P. Hare, C. Webb, & R. Hertz-Lawarowitz (Eds), *Cooperation in education*. Provo, UT: Brigham Young University Press, pp. 14-46.

Smylie, M. A. (1989). Teachers' views of the effectiveness of sources of learning to teach. *The Elementary School Journal, 84*(5), 543-558.

Sparks, D. (1994). A paradigm shift in staff development. *Journal of Staff Development, 15*(4). Retrieved on September 20, 2004 from http://www.nsdc.org/library/publications/jsd/sparks154.cfm

Sun-Keung Pang, N. (2003). Binding forces and teachers' school life: A recursive model. *School Effectiveness and School Improvement, 14*(3), 293 – 320.

U.S. Department of Education. (2001). *No Child Left Behind Act*. Washington, D.C.

U.S. Department of Education. (2006). *Internet access in U.S. public schools and classrooms: 1994 – 2005*. Washington, D.C. Retrieved on May 1, 2007 from http://nces.ed.gov/pubs2007/2007020.pdf

Weick, K. E. (1976). Educational organizations as loosely coupled systems. *Administrative Science Quarterly*, 21, 1-19.

Wilson, S. M., & Berne, J. (1999). Teacher learning and the acquisition of professional knowledge: An examination of research on contemporary professional development. In A. Iran-Nejad, & P.D. Pearson (Eds), *Review of Research in Education*, 24, 173-209.

Wood, F. H., & Thompson, SR. (1980). Guidelines for better staff development. *Educational Leadership, 37*(5), 374-378.

Yocam, K., & Wilmore, F. (1994). *Creating an alternative context for teacher development: ACOT's two-year pilot project*. Cupertino, CA: Apple Computers, Inc. Retrieved on July 12, 2004 from http://www.apple.com/education/K12/leadership/acot/library.html

Zhao, Y., & Cziko, G. A. (2001). Teacher adoption of technology: A perceptual control theory perspective. *Journal of Technology and Teacher Education, 9*(1), 5-30.

Chapter IV

Continuous Practice Improvement

There is no other social issue more critical to the future of this or any other company, the future of the global economy, and the future of our world society than how well we educate our children (Litow, 1999. p. i).

Since the early 1980s education had been challenged to improve student outcomes. It was during these years of debate on what would help schools help children achieve more that microcomputers were making inroads into schools. Apple Computer started the Apple Classrooms of Tomorrow (ACOT) project, described briefly in Chapter III. Microcomputer technology was developing and changing dramatically each year, with new opportunities available to support classroom teaching. Then the World Wide Web came into the mix in 1993, opening up new resources and opportunities for teachers and students. All of a sudden they were not limited to the resources available (or not) in the classrooms or school libraries. The world was opening up to schools in ways never imagined before. However, schools and teachers were still accountable for student learning. How could corporations step up to the challenge of supporting the nations' future workforce?

In 1994, IBM's corporate philanthropy in education announced a new initiative called Reinventing Education (RE). Supported at the highest levels by the Chairman and CEO of IBM, Louis V. Gerstner, Jr., this $35 million grant program arose from the firm belief that business should, and must, contribute to major and long-lasting changes in education. Through RE, IBM brought technology and technology expertise to bear on this process to develop cutting-edge solutions to some of education's toughest problems. The belief was that technology could have the same dramatic influence in public education as it did in business, if the schools had the tools needed for innovative teaching and student achievement. However, the folks at IBM recognized that technology alone was not going to transform complex organizations like schools. "History had already demonstrated that simply donating computers to schools would not raise the level of student performance" (Spielvogel, 2001, p. 3). The process for engaging all stakeholders was emphasized.

Prior to RE, educational corporate philanthropy by technology companies consisted predominantly of donating equipment, funding short lived technology programs, or developing specific, stand-alone products. In all of these cases there was little to no follow-up to measure success or impact, or corporate involvement in the projects.[1] RE moved away from corporate attempts to influence education through one-time grants or awards of technology without support, to a sustained presence and commitment in the process created through long-term educational research and development partnerships between IBM and states or school districts. Rather than impose preconceived biases or solutions to perceived problems in schools, RE targeted locally identified impediments to changes in teaching and learning, and worked with the local educational units to develop interventions based on local needs, while keeping in mind the larger, nationally related issues. To that end, IBM has subsequently awarded Reinventing Education II grants in 1997 and Reinventing Education III grants in 2004.

Selection of initial RE sites was based on specific criteria. "To be selected for RE Grants, districts/applicants were required by IBM to have a comprehensive vision for district-wide restructuring and a track record for innovation and improvement that included a technology plan to help effect a fundamental reshaping of schools" (McLaughlin & Melley, 1999, p. 11). The focus was on creating a professional development reform model that could effect systemic change in the district, not just a single school, and be a national model. In addition, the backing of strong city mayors, personally committed to educational reform, was extremely important. Vital to the grant process was having a commitment from community leaders to ensure success.

At that time, Philadelphia's Superintendent of Schools, David Hornbeck, was implementing what he termed "Children Achieving." Hornbeck believed that all children could learn at high levels and that low achievement was the result of low expectations by those responsible for educating children. Children Achieving was based on 10 points:

1. Set high expectations for everyone
2. Design accurate performance indicators to hold everyone accountable for results
3. Shrink the centralized bureaucracy and let schools make more decisions
4. Provide intensive and sustained professional development to all staff
5. Make sure that all students are ready for school
6. Provide students with the community support and services they need to succeed in school
7. Provide up-to-date technology and instructional materials
8. Engage the public in shaping, understanding, supporting, and participating in reform
9. Ensure adequate resources and use them effectively
10. Be prepared to address all of these priorities together and for the long term – starting now (School District of Philadelphia, n.d.)

Philadelphia mayor, Edward Rendell, fully supported the Children Achieving mission (Kanter, 1999). These two factors, along with the School District of Philadelphia's track record for technology innovation in schools and technology professional development for teachers, helped Philadelphia win one of the first 10 RE partnership grants awarded in 1995, one which proved to be one of the most ambitious of those initial 10. "The IBM grant provided funding in the form of technical resources and cash to develop a professional development delivery system called Continuous Practice Improvement (CPI)" (McLaughlin & Melley, 1999).

In March of 1995 the project team began to work. "Over sixty-five IBM and school district resources contributed to the design and development of the Teacher Professional Development Model, called Continuous Practiced Improvement (CPI)" (McLaughlin & Melley, 1999, p. 3). In this process, IBM established a collaborative working arrangement with the School District of Philadelphia. The goal was for sustained and systematic change across the district.

Working together, a thorough Statement of Work was written outlining three major roll-out phases. Phase I consisted of establishing baseline hardware and software needs and providing participants with training on the software provided. Phase II defined the CPI model, and Phase III was the initial implementation of the model. This chapter describes the development of the CPI model, the implementation of CPI in one of 22 clusters[2] of schools in Philadelphia, and the expansion of that program district wide.

Development of the CPI Model

The CPI professional development model was developed and implemented in an area in north Philadelphia. This cluster of schools was selected for its multicultural aspect and diverse student population. Three schools were initially selected for participation, one elementary, one middle, and one high school. Classrooms in each school were equipped with wiring, hardware and software. In addition, Novell servers were provided to allow sharing of software between classrooms. These classrooms were designated "Exemplary Site Classrooms" and were outfitted with six IBM Pentium1 computers, with one as the teacher station that also had a 17-inch monitor, and a dot matrix printer. Each school received a laser printer, software for the server and classroom computers, and a scanner. All the computers were networked to the server to access the shared software. In addition, these schools were the first to have access to the Internet, through what was called City Net at the time. Internet access allowed the teachers to access the CPI Web site, initially called the CPI Knowledge Site or Wired for Learning,[3] to gain access to the various resources available, along with the journaling areas called CPI Talk and the lesson plan sharing utility (D. Kipp, personal communication, September 14, 2004).

Teachers had to apply to participate in the project. The invitation notice that went out to the teachers in these three schools indicated the project was looking for teachers of students with "special needs" who were interested in improving their teaching practice to help their students achieve more. "Special needs" was interpreted as special education students and English language learners, or ELL students. To that end, the focus of CPI at the elementary school was reading and the development of a program called "WatchMe!Read" for all students, while the focus at the high school was on disabled students and the use of a program called WYNN™ (What You Need Now). Both of these programs were being researched by educational psychologists Dr. Bruce Mahaffey and Dr. Don Nix, IBM research scientists at that time. The intent of the original CPI project was indeed reinventing education through staff development, of which computers were a part but not the focus. Best practices and good teaching were the foci, with using computers as part of instruction being part of the process of being a good teacher. In the end, nine teachers, three from each of the three schools, were selected to be participants in the original CPI project (D. Kipp, personal communication, September 14, 2004).

In the design phase of the model, the nine teachers would come together with the project team from IBM to talk about the project and what needed to be done to design a professional development program. They decided, in consultation with the IBM project team, that they needed to see a program similar to what IBM was suggesting. IBM arranged for teachers and cluster representatives to visit such classrooms in New York City through the Professional Development Laboratory developed by Marion Walsh. Teachers and cluster representatives visited three elementary schools and observed several teachers who participated in the six week program.

These visits happened in June of 1995. A teacher who wanted to improve his/her teaching, or to learn new ways of teaching, applied for the program. If accepted, he/she was paired with a "best practices" teacher. A substitute teacher was secured for the visiting teacher's classroom. This substitute teacher was with the Visiting Teacher in his/her classroom the first two days of the first week to learn the students, the routine and curriculum for the classroom. For the next five weeks, the Visiting Teacher was with the "best practices" teacher observing and interacting with that classroom. At the end of the five weeks, the Visiting Teacher returned to his/her classroom, again with the substitute teacher still in charge for three more days. The intent was for the Visiting Teacher to thoughtfully plan how he/she would implement new skills and teaching practices into the classroom. With this professional development model in mind, the Philadelphia teachers and cluster leaders would begin to develop the Continuous Practice Improvement Model (D. Kipp, personal communication, September 14, 2004).

In planning the model, these teachers, administrators from their schools and the cluster office, working with an IBM team, decided first, and foremost, that a six-week model would not work in Philadelphia, but they liked the overall concept of the Professional Development Laboratory model. To that end, they adopted a teacher-to-teacher, laboratory training methodology similar to "teaching hospitals" rather than the traditional classroom training model. As in teaching hospitals, Visiting Teachers "would be able to strengthen and build upon specific skills by observing and learning from more experienced … teachers" (Coggins, Melley, & Suh, 2000). These experienced teachers were named Resident Teachers, akin to the resident doctors in the teaching hospital. In this model, the Visiting Teachers would work with the Resident Teachers in real classrooms (e.g. "exemplary site classrooms"), where they would see and try out technology enhanced instructional methods in a supported environment. The Visiting Teachers would continue to be in contact with the Resident Teachers even after the Visiting Teachers returned to their own classrooms. They could also continue to have contact with other Visiting Teachers, all facilitated through an online environment developed in conjunction with IBM called the CPI Knowledge Site or Wired for Learning. Through this Web site, connections were available "between pilot schools and teachers to provide on-going support and contact among teachers even after they returned to their home schools" (McLaughlin & Melley, 1999, p. 11).

The fundamental goal of the CPI Model was to improve teaching practice. "CPI achieves this goal through a combination of classroom rotations, technology training, and by fostering an on-going, collaborative relationship between visiting and resident teachers" (Spielvogel, Brunner, Pasnik, Keane, Friedman, Jeffers, John, & Hermos, 2001, p. 36). Within this process, the schools developed a cadre of teachers who collaborated to develop and share teaching practices – teachers were no longer isolated in their classrooms. Teachers saw other teachers using computers and wanted the same for their classrooms.

What was truly ambitious was taking teachers who had no previous experience using computers in their classrooms, providing them with sufficient equipment and training, and asking them to demonstrate to others what they had learned. Establishing these exemplary practice schools and classrooms required a serious level of commitment by the teachers selected and principals who supported them. Through the initial phases of the CPI experiment, the program depended on principals to coordinate Visiting Teachers' access to the Resident Teachers. It quickly became clear that the process could not be totally dependent upon principals to make this happen and there needed to be a program coordinator or director to oversee all aspects of the program across all schools. However, it should be noted, as it was in Chapter I, that the overall support by principals for technology in schools was, and continues to be, a primary ingredient for success. As will become evident in the discussion of the teachers' experiences with CPI, the principals' encouragement and willingness to provide opportunities for the teachers to learn and practice new skills in teaching with microcomputers and technology was invaluable.

Initial Implementation of the CPI Model

The original 22 CPI grant teachers were identified by the School District of Philadelphia through the process described above, and their classrooms were outfitted with hardware and software. IBM specialists trained these teachers on the hardware (e.g., the Novell server and the Windows 95 operating system on the IBM computers) and software that was installed on the server and the computers in their classrooms. None of the teachers selected used computers in their classrooms prior to the grant. In fact, most of these teachers had no or minimal experience at all with computers. When the equipment was installed in the classrooms, IBM provided instruction on Saturdays and during Professional Development days in WatchMe!Read at the elementary school, What You Need Now (WYNN) at the high school, Internet and e-mail, Lotus Notes®, IBM School Vista, along with select software to support reading, writing, science, and math curricula. The intention was not that they would necessarily use all of these programs, but they would have access to them as needed.

Eventually, only nine teachers were selected from the original 22 to be Resident Teachers. These nine teachers became the key facilitators in development and refinement of the program. One of these teachers received training in the concepts of Critical Friends Groups (CGF) by the Coalition of Essential Schools and became certified as a coach. Not all the Resident Teachers participated in the CGF, but used a less formal approach to communication to sustain relationships after visitations. Since many initial Visiting Teachers were from the same schools as the Resident Teachers, ongoing communication was easy to accomplish without technology mediation. They could meet before school, over lunch, and some even carpooled together to work where they could discuss ideas.

CPI worked on the principle that teachers can visit an "exemplary" classroom in their own school or one nearby to observe and learn new and innovative teaching practices and how to infuse microcomputers into their own classrooms. Ideally, a first grade teacher would observe another first grade teacher, a third grade teacher would observe another third grade teacher, and so forth. With only one exemplary classroom in a school, visiting teachers were more likely to observe a Resident Teacher who taught a similar grade level and classroom environment. Over time, with addition of new Resident Teachers, the pool of options expanded substantially, but initially there were limits.

The Model

Here is how the model works. There were five roles to be played out. As described above, there were Visiting and Resident Teachers. In the initial roll-out of the CPI Model, these Visiting and Resident Teachers were all within the north Philadelphia area. To allow the Visiting Teachers to leave their classrooms for up to three days for their rotation visits, a Substitute Teacher was identified. This person was a well-trained substitute teacher who filled in for the Visiting Teachers so the students in their classrooms moved forward and the school was not burdened with a classroom without a teacher. A full time Substitute Teacher was hired in 1997, but resigned in 1999. There were Mobile Facilitators, or Educational Technologists, identified to support the process from the cluster level. And lastly, the District identified a CPI Project Director from the area who oversaw the process of training and rotation assignments, and monitored the online discussion.

The initial CPI "Wired for Learning" Web site was available for collaboration with other CPI teachers and access to resources. This Web site was available 24 hours a day, seven days a week for the CPI teachers to communicate with one another or to write journal entries about their successes and problems, thoughts and concerns, and to post exemplary lesson plans to share with other CPI teachers. Only CPI teachers could gain access to this resource which made it a special site for these teachers. Both the Visiting and Resident Teachers were asked to make journal entries during the classroom visit rotation and afterward to document and keep track of questions, issues, successes, and more. They could return to these postings at any time to revisit and to update progress. In addition, IBM paid for the Achievement Standards and the Curriculum Frameworks of the School District of Philadelphia to be available in Wired for Learning. Only the CPI teachers had access to this resource online. While this may seem trivial to some, in the late 1990s these resources were only accessible otherwise via a CD-ROM available from the school district or in print. Often classroom teachers did not have their own copy to use.

Before any rotations started in 1997, the Resident Teachers observed each other and then came together to talk about the process. A selection process took place, some based on self selection and some based on skills, that produced only three final Resident Teachers, one each at the elementary school, the middle school, and the high school. One example of the selection process had to do with one of the original Resident Teachers who was quite "computer phobic." I'll call her Jane. If there were a Visiting Teacher who was deemed to be, or considered himself to be, "computer phobic," they would assign that teacher to observe Jane. The theory was: if Jane could get over her fears of computers and get students into using computers as part of her teaching, anyone could. Most of the Visiting Teachers went to the elementary or middle school, rather than the high school (D. Kipp, personal communication, September 14, 2004).

In addition to the CPI Model of visiting Resident Teachers during the school year, a summer institute was developed through the school district in collaboration with the elementary and middle schools. Only teachers from the area schools were eligible to participate. Teachers from the first expansion middle school could apply initially for this summer program, just as they had for the regular program. Eighteen teachers participated. The announcement gave the following description of the CPI program:

Continuous Practice Improvement is an approach to professional development that allows teachers-to-teacher collaboration within actual classrooms. A Visiting Teacher makes an extended visit to a Resident Teacher's classroom to work on professional goals and to further the Visiting Teacher's ability to integrate technology tools into the instructional program (CPI Summer Institute Flyer, n.d.).

Teachers spent one of three weeks during summer school observing at either the elementary or middle school. At the elementary school the focus was on reading and middle school teachers collaborated on a food safety project. Teachers who participated in this summer program were expected to have at least one computer in their classroom already that was connected to the World Wide Web. In addition to the personal mentoring experience, these teachers were compensated for the four hours per day at the staff development rate in the district at that time. Not just anyone could apply. Applicants needed to have taught for at least two years full time in the district with a satisfactory teaching record.

Seventy teachers successfully completed their visitation rotations with Resident Teachers from fall 1997 through the 1998 calendar year, and 143 teachers completed rotations through Spring 2000 with an estimated 4,700 students impacted. This created a cadre of teachers within the area and a momentum toward establishing changes in classroom practice and student achievement. The number of teachers involved could have been higher except for the problem of identifying substitute teachers. This became a perennial problem for the CPI program over the years it was in place in the school district.

One exceptional aspect of the original CPI project was how the Visiting and Resident Teachers bonded (D. Kipp, personal communication, September 14, 2004). Through the process of supporting best practices in teaching, continued conversation between teachers, and a sense of ownership of their professional development, these teachers developed a "community of teachers" to develop and share practices. This "community of teachers" best fits the definition proposed by Grossman and Wineburg (2000), as opposed to a gathering of teachers, in that they shared a common set of experiences and were intent on sharing practices to support each others' endeavors to improve the learning environments for their students.

As one can see, the emphasis was on teacher-to-teacher professional development within an actual, functioning, successful classroom. This connection, coupled with a rich, ongoing, online support mechanism, provided teachers just learning how to manage a classroom with microcomputers with resources to help them through the process of changing practice. This model of professional development was considered potential for the following benefits:

1. Increased student achievement
2. Increased teacher expertise
3. Increased student and teacher access to technology
4. Enriched and expanded knowledge Web site
5. Expanded teacher networks and professional relationships
6. Improved quality of professional development for teachers
7. Opportunity to test instructional program effectiveness
8. Increased job satisfaction (McLaughlin and Melley, 1999, p. 14)

Before expansion beyond the initial schools, the CPI program provided the schools within this cluster a cadre of teachers committed to collaboration and support for other teachers. It is important to remember that the CPI program was not about a particular subject or teaching method, but about good practice in classroom management of the teaching and learning enterprise. Through modeling best practices for infusing computers into lessons and activities, these Resident Teachers were able to help the Visiting Teachers understand the power the technology can unleash in the classrooms. The fundamental point was to achieve a well-managed classroom, using the power of all the resources available to support student achievement at and above the school district expectations. With this model, teachers would be able to successfully teach any subject using microcomputers as one teaching tool.

Expansion Beyond the Original Schools

One of the CPI teachers from the first group of schools, Ms. Maguire (nee Moore), saw the potential of this model for expansion across the district. She approached the director of the Technology Learning Group at that time who liked the idea. They approached the principal at a high school in west Philadelphia, who agreed they could provide staff development training within that school. The CPI model was modified such that the teachers were not only visiting other teachers within this high school, but Ms. Maguire worked with teachers one-on-one on how to incorporate computer labs into their curricula—a slight modification. Ms. Maguire worked specifically with two teachers over a few months until they were comfortable using a computer. These two teachers got one computer for each of their classrooms and served as Resident Teachers for the other teachers to visit within this high school. Again, it was a "local model," in that the intent was for teachers to visit other teachers within their school, if possible. Fourteen teachers were trained altogether and Ms. Maguire was at the high school two and one-half days a week for a year. She spent time bringing students into the computer labs to show teachers how to use the labs with their students. It was not so much about integrating computers into the small classroom, but how to use the mini labs—community labs with about 12 computers—to support teaching and learning. Ms. Maguire's responsibility was to work with teachers on curriculum-oriented applications of computer labs, as well as the software available there, to support the high school curriculum (D. Maguire, personal communication, October 26, 2004).

These teachers signed up because they were interested in using computers and the program was during the school day, not after school or on Saturdays; therefore, it did not add to their work day or take away from weekend time (D. Maguire, personal communication, October 26, 2004). While this may not seem important in the long run, the CPI program was just evolving. As funding sources became available to support the program, time for training in uses of microcomputers and the Internet became essential components, meaning teachers would need to take personal time to attend the training.

IBM recognized the program at the high school as a local implementation of the CPI model with important local customizations. Given the success of CPI in the two experiments, the School District of Philadelphia included CPI in a Technology Literacy Challenge Fund (TLCF) grant application in 2000 as one of three technology related professional development opportunities for teachers. While the emphasis remained on enhancing teaching as a whole, infusing computer technology into the teaching and learning process became equally important. Ms. Maguire continued as the district wide CPI director who oversaw all applications by potential Visiting Teachers, identification of new Resident Teachers and additional Substitute Teachers, conducted most of the training of Visiting Teachers, and was available to the Visiting Teachers in their schools in times of crisis. As will become evident, having a program director who is steeped in the program and dedicated to the model was extremely important. She would visit the Visiting Teachers' classrooms to provide support, identify additional resources in their area, as well as provide technical guidance if needed. As the person who provided most of the training sessions, she came to know each and every Visiting Teacher extremely well. These relationships and her intimate knowledge of the classrooms gave Ms. Maguire a unique opportunity to identify new Resident Teachers from past Visiting Teachers (Schifter, 2006).

The TLCF grant included incentives for teachers to take the courses. Teachers who completed any of the three professional development courses, including CPI, received a desktop computer, a printer and five age-appropriate educational software packages (including Microsoft Office Suite and Inspiration) for their classroom. The 30 hours of training was applicable to the Commonwealth of Pennsylvania's new Act 48 credits.[4] One new major change in the CPI program was introduction of Apple computers, since the district had adopted them as their standard for the elementary and middle schools. Thus, the desktop computer they received for their classrooms was an Apple computer. (In the original CPI Model, supported by IBM, the teachers and schools received IBM computers. In 2005, the School District of Philadelphia replaced the remaining non-Apple computers in those schools with Apple computers for all elementary and middle schools to be using the same platform.)

CPI is a program designed to introduce teachers to resources that help them integrate technology as a cross-cutting competency across the curriculum within our classrooms. This course is aligned with the curriculum frameworks and can be utilized in a classroom with as little as one computer. In addition, it will model effective teaching practices using a variety of best practice strategies. Components of this program include a two-day observation visit to a classroom where technology integration is implemented on a daily basis, along with on-line collaboration and support (Moore, n.d.).

This notice went out to all elementary and middle schools to be made available to all teachers in those schools. Only 15 to 20 teachers were accepted into any one CPI cohort, and for the first two years there were two cohorts at a time. The aim of the program was to train at least 110 teachers each year, thus serving approximately 6,000 students across the school district. In the long run, the CPI program had five goals: to improve student achievement, improve teacher expertise (in using computers), increase access to computers, expand professional relationships, and improve professional development related to computers in education. The rest of this chapter will present the CPI program in its final form given the advice of CPI Visiting Teachers along the way, including a description of the overall program, followed by detailed information about each training day in the computer lab, what was expected between training dates, and the process of the classroom visitation rotations.

TLCF CPI Program

The TLCF CPI visitation or rotation process was similar to the original CPI program. This included for the Visiting Teacher day one with the Substitute Teacher, the next three days with the Resident Teacher, and the last day back in their classroom with the Substitute Teacher still in charge. In addition to continued access to the Resident Teacher, the cluster Educational Technologist (ET) would visit the Visiting Teachers the first and fifth days, and the school Technology Teacher Leader[5] would be available as needed.

During the time of the first group of teachers participating in the TLCF CPI program, Ms. Maguire noticed that these teachers did not have basic computer knowledge and proposed including a course on computer basics (e.g., how to turn on the computer, how to set up student folders, how to manage files, how to use e-mail, among other skills) (D. Maguire, personal communication, October 26, 2004). These teachers were receiving a new computer for their classroom, and many were not comfortable even turning on a computer. This course was developed initially for 15 hours of time, which included 12 hours of in lab time over three Saturdays within a five-week period, where they learned how a computer works, MAC OS 9, Internet and Web browser software basics, along with training in e-mail and several educational software applications. In addition there were initially three hours of homework, or one hour of homework for each four hour class. The homework consisted of three reflections—one before they started the program, one after the observations, and one after the program was completed. All of this was in addition to their visiting a Resident Teacher for three days during a school week. The course evolved into a 30-hour program including 15 hours of hands-on training, 15 hours of independent work (e.g., practicing using software, writing journal entries and responding to others in their CPI cohort/class), and the three days with the Resident Teacher.

The new computer was delivered to the teachers on the first Saturday of the CPI course. The computer was actually assigned to the teacher's school, but specifically for this teacher's use. If the teacher transferred to another school, the computer remained with the original school. Teachers reported to Ms. Maguire that they wanted to practice what they were learning and do the homework, but they were limited to only being able to access the computers at school, during times when they were not actively teaching which took away from preparation time. Ms. Maguire then proposed for the grant to include a laptop loaner program in conjunction with the desktop computer to begin in Spring 2001. The laptop, as a loan only for while they took the course, was to be taken home by the teacher for practice using the new software applications, accessing the Internet for resources, and for doing the homework through the CPI Web site. The laptop loaner was returned after the teacher completed all components of the course and rotations. The desktop computer remained at school.

The Saturday dates for laboratory training were staggered such that an introductory class would meet the first and third Saturdays in a month followed up with the second Saturday of the following month. In the intervening Saturdays an intermediate group would have training.[6] After six weeks of training, two classes or cohorts of CPI Visiting Teachers would have completed the laboratory and online portions of CPI and be ready for classroom visitations. In year three, or 2002, the program experimented with two online sections for more advanced users of computers and three sections situated out in Community Technology Education Centers located in schools strategically positioned across the district (one in south, one in north, and one in northeast Philadelphia). Both of these efforts addressed feedback from previous CPI teachers that the location of the training was not convenient.[7] The online component did not require the teachers to physically come together to learn about the programs, and emphasized the interaction through the CPI Learning Village™, formerly known as Wired for Learning, online site. Teachers who participated in the online program were selected because they had already demonstrated through other professional development courses that they had more than a basic understanding of computers. In the end, after five years of offering the program under the direction of Ms. Maguire, 294 teachers (29 males, or 9 percent, and 265 females, or 91 percent) from 63 schools completed the program.

The one prerequisite was "basic computer knowledge," which was assessed using the LoTi Survey, or Levels of Technology Implementation Survey by Dr. Chris Moersch[8], which measures the level of technology use in classrooms. Unfortunately, this survey is a self-report survey, meaning there were teachers accepted into the program who did not report "basic computer knowledge," including skills with e-mail and some productivity software applications. This reality forced the program to address the teachers where they were in their technical capabilities, truly a differentiated learning model. The director emphasized that the program was not intended for everyone to "graduate" with the same knowledge set, but each participant would develop their classroom computer acumen as appropriate for themselves. Thus, some teachers started out with very little to no knowledge or skills and completed the program just being "comfortable" with their use of computing with their students; other teachers started out with greater knowledge and skills in using computers and completed the program knowing how to use computers more efficiently and effectively within the curriculum and with their students. While addressed, the emphasis of the CPI program was not in how to use computers to teach literacy or mathematics or any specific subject, but accentuated how to manage a classroom and set of activities so to maximize resources available, of which computers were a part.

In the Laboratory

At the first session, the teachers were asked to complete a short survey assessing their knowledge and comfort levels with computers. This survey was different from the LoTi survey because it was not asking about computer use in classrooms. It was only intended to learn how these teachers perceived their knowledge and comfort with computers. They were again asked to complete the same survey after completing their CPI training to assess change in their reported perception of knowledge and comfort with computers. This information showed that more than 90 percent of the CPI participants were comfortable or very comfortable with using computers by the end of the training. In particular, several computer applications registered excellent improvement in the percentage of teachers who were comfortable using them, as follows: use and creation of slideshows (58.6 percent increase, from 23.9 percent to 82.5 percent); creating or modifying Web pages (48.9 percent increase, from 9 percent to 57.9 percent); strategies for classroom computer use (43.3 percent increase, from 26.9 percent to 70.2 percent); and knowledge and use of Inspiration™ software (43.2 percent increase, from 16.4 percent to 59.6 percent). Increase in knowledge and comfort about using the Internet was more modest (only 10.6 percent), but the starting and ending points were much higher (from 80.6 percent to 91.2 percent). Relatively high starting points were also reported for the percentage of teachers reporting comfort in using browsers, search engines, and bookmarks (65.7 percent); and locating resources for classroom projects (59.7 percent). For these two items, increase in the percentage of teachers who were comfortable or expert by the end of the training ranged around 20 percent (Vilceanu and Schifter, 2002).

Using a PowerPoint presentation, the instructor reviewed the CPI professional development model. From the beginning, the instructor set the stage by telling the teachers what they already knew, but probably had not recognized: "Teachers learn best by directly observing and experiencing the application of technology in the classroom with hands-on experience, working with students using technology, collaborating with the Resident Teacher, and reflecting on practice" (Moore, 10/3/00, PowerPoint slide #4).

One metaphor explored to explain how the CPI program works was the Native American Medicine Wheel. "A Medicine Wheel is a physical manifestation of Spiritual energy. An outward expression of an internal dialogue. A mirror in which we can better SEE what is going on within us. It is a wheel of protection and enables us, and allows us, to gather surrounding energies into a focal point and to commune with Spirit, Self and Nature (ALL elemental forces)........Creation!" (Tree-Song, n.d.). The Medicine Wheel is an attempt to explain nature and suggests that all creatures are part of the same natural order of things, and, therefore, support one another through life. The strengths of one person are the weaknesses of another; therefore, they are complementary. In using the Medicine Wheel metaphor, the director of the CPI program was indicating to the teachers that it was important not to focus on their own weaknesses or lack of knowledge, but on how teachers could support one another in their goals to support the students' achievements. Where one teacher had knowledge and strength in an area of study, that teacher could support and help teachers who had less knowledge and strength. This symbiotic relationship goes back to the very nature and essence of the CPI program—teachers helping teachers.

The inter-classroom visitations component of the CPI program was further explained in basic terms. On Day 1, the Substitute Teacher[9] would be with the CPI Visiting Teacher to learn about what lessons were planned for the next three days. Days 2, 3, and 4 the Visiting Teacher would visit the assigned Resident Teacher to observe, interact with, and reflect on classroom practices. Then on Day 5, the Visiting Teacher returned to her classroom to plan for integration of technology, coach selected students to be technology gurus[10] for the class, and complete the post-rotation survey about what they learned. The Substitute Teacher was still in charge of the classroom through Day 5.[11]

The group and distance learning coursework represented all the components of the CPI program that prepared the Visiting Teachers for their rotations. The three distance learning sessions focused on Internet research (e.g., learning about Internet portals and kid-safe search engines) and microcomputer-integrated lesson building. These sessions focused on what Ms. Maguire called The Quality Classroom. The four characteristics of The Quality Classroom, as described by Ms. Maguire, were:

1. The LOOK or physical space and management strategies. The teachers were encouraged to look at the arrangement of their classrooms, including where computers were located for accessibility by students and the teacher. The tools to consider were computers, the local area network to the school server, books, learning centers all designed for instruction, assessment, success, and restructuring.
2. The FEEL of class interactions, or the development of a safe environment. The teachers were encouraged to consider student interactions with computers individually for reading and writing, small groups at work stations and in discussions, and whole class presentations or demonstrations. In addition, interaction with other CPI teachers would help teachers expand their repertoire of ideas. As noted by Glasser (1993), the classroom environment should be warm and supportive, not adversarial or coercive.
3. The HEART of the program including content standards, curriculum frameworks, software supports, and using research to inform practice. Here the Philadelphia Curriculum Frameworks[12] were to be used as the facilitator for using the Internet for student research, using word processing and Web pages to support student writing, and supplemental software that they would learn for practicing skills, learning content, and enrichment activities.
4. The ACHIEVEMENTS as measured by presentations, writing, and authentic examples of student work. Demonstrations of understanding, learning and achievement were identified as presentations of work through student exhibits, slide shows, poetry, stories, graphics, and spreadsheets, and based on explor-

ing, planning, and building multimedia electronic portfolios. Overall, the goal was for an integrated creative process.

All of these characteristics would be interwoven through collaboration promoting success, and through management as the tool for teachers and students to know where they want to go and to know when they get there.

Lastly, ongoing collaboration and support were important to the success of the CPI program. The director stressed the importance of the electronic/online conversation facilitated through the CPI Web site, and the follow-up visits provided by the regional Educational Technologist and the local Technology Teacher Leader in the school. By the end of the six sessions, the three on Saturdays and three through distance learning, the CPI Visiting Teachers would be prepared to set goals for themselves to transform their classrooms to emulate The Quality Classroom.

Four categories of classroom use of a single computer were explored (Dockterman, 1998), including one computer as a presentation tool (pp. 43-54); one computer and interactive group software (pp. 55-72); one computer as a learning station (pp. 73-78); and one computer for the teacher's personal productivity (pp. 25-42).

For many teachers, the concept of using a single computer with a class of 25 to 30 students seemed daunting. Teachers became less concerned as they explored these scenarios. If they could imagine how one computer could transform their classroom, the idea of additional computers for learning or technology centers and projects became attainable.

Still on the first day of the CPI program, the class reviewed the basic components of a computer and how a computer functions, for example, which parts are the "brains" of the computer or the "nervous system" of the computer, and how these parts communicate to produce text and images on the computer or to print. All of this was as introduction to the program, setting the stage for learning how to use Eudora® for e-mail that would become a major communication tool for the course, accessing the CPI Learning Village Web site to participate in the on-line discussion board, and basic uses of the Internet and search engines. The teachers were introduced to "kid-friendly" search engines, such as Ask Jeeves for Kids® and Yahooligans®.

During the last two hours of the day, the teachers were issued their loaner laptops and were given time to explore how to use the laptop. The laptops were Apple i-Books, so the instructor went over the basics of the Mac Operating System, which at that time was OS-9. She taught them how to configure the laptop to be used with a phone line to dial into either the CPI Web site, or the school district to access e-mail using Eudora e-mail software. For many of these teachers, accessing e-mail was a unique experience. While all teachers in the district had been issued e-mail accounts in the mid-1990s, many teachers did not use them, so passwords no longer were remembered nor worked.

They were given written instructions to take with them, but the practice time was most important to give time for questions, concerns to be addressed, and comfort to be established. Teachers would help each other. As teachers were successful in connecting and making the change from one configuration to the next, they would help their neighbor who might be less confident.

The last component of the first day was introduction to the CPI Learning Village Web site where the teachers were introduced to the online discussion board and encouraged to answer the first writing/journal prompt, or Activity 1, to introduce themselves to the group, including what grade they taught, something about the school where they taught, and what they hoped to learn through the CPI program. In addition, they were asked to identify three new things they had learned that day, two questions they had after the session, and one idea they will implement right away in their classroom.

On the second Saturday session there was discussion about the quality of the Resident Teacher classroom, what the teachers could expect to learn from the Resident Teachers. This discussion was important since these teachers needed to begin to give the CPI program director indications about when they might be available to complete their residency, again following the medical model as described in above for the rotation component so she could make appropriate matches between the CPI Visiting Teachers and the Resident Teachers. It was during this second session that the teachers were introduced to creating a slide show, creating a Web site using the online Web site tools, and learning the concept-mapping software Inspiration. During the 5 hours of the session, each new program was given approximately 1 1/5 hours of instruction and practice time. The teachers were expected to practice these new skills over the next two weeks before the final Saturday session. In addition, the second activity through the CPI Web site was to post a journal statement discussing one possible application of the Internet search engines in the classroom to share ideas with each other and to build a community of CPI teachers who could support one another over time. Lastly, they discussed the issues of student use of the Internet, including Internet Safety and what strategies they might use to protect their students. One such strategy was to provide safe entry ways to the Internet, such as Ask Jeeves for Kids or Yahooligans, or to bookmark sites specific for assignments. For these teachers, controlling access and search options was very important to provide a "safe" classroom for all.

During the third and final Saturday session, the teachers worked on the Instructional Planner component of the program. The program director introduced what she called the Seven Steps to Success: A Model Lesson Plan (Moore, 2001). These steps were to gain attention of learner, review relevant past learning, communicate the goal of the lesson, presentation of new material, provide guided practice, provide independent practice, and close the lesson. These are quite similar to Slavin's Quality of Instruction model that purports to state the learning objectives to orient the learner, review the prerequisites for the lesson (prior knowledge), present new materials, probe for learning, provide independent time for practice, assess performance, and provide feedback, and assign homework to further hone the knowledge/skill (Slavin, 1997, pp. 237 - 248).

Next, the teachers worked in grade teams to create lesson plans using the seven step model to use with their classes. By the end of this session, the teachers would have one complete lesson using the Inspiration concept mapping software and a complete second technology integrated lesson. By this session they should have completed a slideshow to be used in their classes, and they would share this slideshow with their CPI classmates. In addition to these activities, the group discussed how to identify and train students to be technology "experts" and how to use these students to support teaching and learning for all their students. If there were a computer problem while a student was using the computer, the student was to hail one of the student "experts" before contacting the teacher. If the student expert could not solve the problem, then the teacher would be summoned, but only as a last resort. This process would allow the teacher to continue teaching a lesson to a subset of the class rather than being constantly on-call for technology issues, and would give a few students in the classroom responsibility to help classmates in trouble. In the end, this last session was a time to celebrate all that these teachers had accomplished in five weeks, and to prepare them for their Resident Teacher rotations.

To help the teachers accomplish all their assignments and goals, extensive support materials were created. There was a training manual that was a loose-leaf binder into which teachers could place any and all handouts to save and reuse as needed. They were given explicit instructions. Nothing was left to chance regarding teachers taking notes where they might miss a point or not jot down something important.

Between Laboratory Sessions

Between each session there were readings posted to the CPI Learning Village Web site. These readings were about current issues in technology integration, sometimes controversial. The teachers were expected to explore different search engines and portals to be used either by themselves or by their students, and to gain an understanding of Web site evaluation guidelines for children. In addition, they were asked to share their experiences and opinions with their fellow CPI classmates.

Also between sessions, the CPI teachers were asked to post entries to their CPI journal within the Learning Village site. One specific question they were to answer each session was "What did you learn today and how do you expect to use this new knowledge in your classroom?" During their rotations with their Resident Teacher, the CPI teachers were asked to post to this journal what they learned and how they planned to change their own classroom. In addition, the Resident Teachers were encouraged also to post to the Learning Village site to share their experiences and insights with other CPI teachers.

These journal entries were analyzed qualitatively to determine whether there were any themes that arose from their discussion that should be known and possibly addressed by the program. Five themes arose from the data, presented in alphabetical order and not priority order: frustration, learning something new, new ideas to try, satisfaction, and watching the kids (Schifter, 2001). Frustration arose over not finding assignments, limited Internet access at their school, concerns about their rotation assignment, and limitation of time. These frustrations came out of not being in control of their learning and situation, and the fear of the unknown. Many schools in Philadelphia in the late 20th and early 21st centuries still had limited technical infrastructure, such as few electrical outlets in classrooms and Internet access only in computer labs. These problems were not under the control of the teachers, but caused consternation due to not being able to accomplish goals they set through the program. Learning something new was evident by statements about learning new ways to engage their students through presentations, slide shows, Inspiration, the Internet, learning stations, and more. They spoke of learning how to use the computer as both a teaching aid and as a learning aid, neither of which they had considered before. As they observed students seamlessly using computers to accomplish learning objectives, they indicated their amazement and the possibilities for their classrooms. Similarly, the theme of "new ideas to try" was exemplified by comments regarding new techniques for infusing computers into lesson planning and the classroom as a whole.

Satisfaction was about the program overall, the organization of the program, the quality of the rotations, how students had responded to the new computer access, and success in using the computer with their students. There were accolades for Ms. Maguire and the other instructors who assisted Ms. Maguire on occasion, and praising the program for treating them as professionals. Lastly, the theme of "watching the kids" came from comments about the teachers' enjoyment of seeing how the students engaged with the computer in accomplishing goals to support learning across subject areas. While there were some frustrations along the way for some teachers, the vast majority of comments teachers made through their journal entries were about what they were learning and observing, and the impact of the program on their classrooms.

The Resident Teacher Rotation

Each CPI Visiting Teacher was assigned to a Resident Teacher who had been identified by the school district as being highly competent in infusing computer technology into the curriculum, and by Ms. Maguire as being able to communicate his/her knowledge to another teacher effectively. The intent was to have identical grade assignments, so that a third grade CPI Visiting Teacher would be with a third grade Resident Teacher. This didn't always work out, but typically in these cases the

Resident Teacher was from only one or two grades different from the CPI teacher. In order to maximize the use of Substitute Teachers, rotation assignments often were staggered over a period of four to six weeks. This would allow the same Substitute Teacher to work with four to six Visiting Teachers, which could mean a CPI Visiting Teacher might have to wait several weeks to conduct their rotations. If this were the case, the CPI Visiting Teacher was encouraged to continue to practice the skills they had learned so far, to establish the new desktop computer in the classroom using principles learned through the CPI program, and to attempt to incorporate the skills and new computer into current classroom activities. The process initially was for five days, but was revised after the first year to four days, as follows:

- **Day 1:** On this day the Substitute Teacher would spend time with the CPI Visiting Teacher getting to know the environment and protocols of the CPI teacher's classroom, meeting the students, and getting information about the curriculum to be followed while the CPI teacher was away. The cluster educational technologist would review technology basics from the program, answer any questions about the new desktop computer, make sure the desktop computer was installed and configured appropriately, and answer any questions for the CPI Visiting Teacher in preparation for their observation days.
- **Day 2, 3, and 4 (later only 2 days):** On these days, the CPI Visiting Teacher visited the classroom of their assigned Resident Teacher. The Visiting Teacher's role was to observe every aspect of the classroom routine and take extensive notes about what they saw and learned. In some cases the students in these classrooms demonstrated for the CPI Visiting Teacher how they made use of computers in their work. In some cases the Resident Teachers provided resources for the CPI teacher to take with them. In one school there were several Resident Teachers in different grades. These teachers would ensure that the Visiting Teachers in their classrooms also observed activities in other Resident Teachers classrooms. As one of them noted, "you are not teaching using technology all the time, so why not have the teacher visit another classroom while you are at lunch or recess rather than observe an irrelevant activity" (anonymous, personal communication, November 3, 2004).
- **Day 5:** The CPI Visiting Teacher went back to their regular classroom, with the Substitute Teacher still in charge. The CPI teacher was to plan how they were going to prepare their classroom for technology integration—computer rotation strategies, establishing a technology center, aligning new computer software with the curriculum, and possibly physically rearranging the furniture in the classroom. They were asked to revisit the four characteristics of The Quality Classroom and envision what changes were needed in their classroom to meet those characteristics. The cluster educational technologist was available, or on call, to assist if needed at this time.

In addition, the CPI teacher still had access to the Resident Teacher through e-mail to clarify points, to seek advice, and to share success stories. This connection between the Visiting and Resident Teachers often continued for some time after completing the CPI program, giving the CPI Visiting Teacher a level of support not available through other technology-related professional development programs available through the district. In addition, there was always the CPI Learning Village site to both post comments and success stories to other teachers from the program, and to glean ideas for lessons and activities from all the postings from other CPI teachers. This resource was available to all CPI teachers until it was discontinued in 2005.

Follow-Up Activities

Approximately one month after these teachers completed the full CPI program, the trainings both face-to-face and online, and the rotation with the Resident Teacher, the cluster educational technologist would visit their classrooms to offer further demonstration lessons, technology support, and assistance with lesson planning and implementation. In some cases, the Resident Teacher was in the same school as the CPI Visiting Teacher. This was a fabulous arrangement because the CPI Visiting Teacher could continue the discussion with the Resident Teacher quite easily without having to use e-mail or have the time lapse between asking a question and receiving the response. The Resident Teacher could provide additional local support long after the observation period. When the CPI Visiting Teacher and Resident Teacher were not from the same school, the program encouraged them to stay in touch through e-mail or the CPI Web site and discussion board. Often CPI teachers would post "aha" instances to the CPI Web site offering lessons they had tried that they believed worked beautifully with their children. This sharing further supported development of the community of teachers that the CPI teachers became.

In the 2000-2001 was very aggressive in recruitment, going after each cluster of the district, and the director had access to the educational technologists (ETs) in each cluster who were supporting the program. The ETs visited each new Visiting Teacher on or before "day one" of the CPI rotation week, and again when the Visiting Teacher finished the rotation. On the first day the ET worked with the new Visiting Teacher for one half day to make sure the new computer was set up properly and to talk about what they should concentrate on in the observation of the Resident Teacher. On the last day the ET was to make sure the Visiting Teacher knew what he/she was doing and spent another half day answering questions and troubleshooting any problems encountered. One month following completion of all aspects of the CPI program, the local cluster educational technologist would visit the CPI teachers in their region to offer further demonstration of CPI principles, technology support, and assistance with lesson planning and implementation. This process was very time intensive.

Then the district structure changed. In 2000-2001 there were 22 clusters of schools. Reorganization of the district resulted in only nine regions, which resulted in fewer regional ETs and the remaining ETs having responsibility to work with many more schools than before. From this outcome, the regional ETs could no longer support the program as they had before.

During the 2001-2002 school year, the same procedures were followed. While there was a great deal of interest by teachers to sign up, Ms. Maguire could not secure enough Substitute Teachers to accommodate visitation rotations for all the interested teachers; therefore, fewer teachers completed the program. Ms. Maguire reported that it was always easy to get substitutes into some areas of the city. "People simply didn't want to go into the really low achieving areas of the city. And when they did go, issues arose. One substitute walked out of a school in mid-day because it was too rough" (D. Maguire, personal communication, October 26, 2004). Substitutes could sometimes be placed in the "harder areas"[13] of the city for Kindergarten through grade three, but not above grade three. In addition, the program cost $494 per teacher, which included the computer and the cost of the substitute. The cost became a factor when decision makers saw teachers could take staff development courses through the Community Technology Education Centers (i.e., CTEC, the nine regional educational technology centers in Philadelphia available through schools to the surrounding community for training) and realize more teachers trained for the same funding.

One outcome of the CPI program was relationships established between teachers. While this was similar to what was originally developed as a community of teachers, there was more. Ms. Maguire had events at Christmas and the end of the year to celebrate teachers' successes. People came out to these events because of these relationships developed through CPI across the district. "There was a sense of support, like a mother – daughter thing. There are still connections." (D. Maguire, personal communication, October 26, 2004) In many ways, the "mother" was Ms. Maguire. She cared for these teachers, provided nurture and support, much like a mother does. This is not something to take lightly.

When Ms. Maguire selected Resident Teachers for the CPI program, there were criteria that were not just about the use of technology in the classroom. Could these teachers relate to other teachers? "They had to be somewhat extroverted. They needed to be willing to share. When there was an RT who was not willing to share or have the people skills, they were not assigned VTs." (D. Maguire, personal communication, October 26, 2004) This was important, since it was not just about technology. The Resident Teacher's job was to demonstrate to the Visiting Teacher what he/she did to infuse microcomputers and other technology into lessons and learning activities. Being willing to share with other teachers meant being able to verbalize their aims, intentions, practices, policies, and more. Communication was very important, just as in teaching in general.

Another success was the course itself. At each of the Saturday sessions, the group of Visiting Teachers got to know each other well. They learned that not everyone needed to learn at the same pace or be at the same level of computer comfort or use as everyone else in the group. "If you started out at a level of one, or basically no skills, you were only expected to reach up to a level three, and that was okay. They realized from day one no stress was put on the group for all to be at the same level. They were told that absolutely not everyone is going to have the same skill levels at the end, but that is okay" (D. Maguire, personal communication, October 26, 2004).

After the initial TLCF grant was finished, feedback from Visiting Teachers questioned the need for so many days of observation during the rotation period. The CPI Director considered this aspect, and reduced the number of days of the visiting rotation to two days for the second TLCF grant. Feedback from the Visiting Teachers during this grant period indicated teachers were satisfied with the amount of time observing a Resident Teacher and did not believe more time was needed.

Resident Teachers

In discussing the CPI program, the Resident Teachers need to be discussed. As noted above, they were the resource teachers whom the Visiting Teachers observed using computers in classrooms for teaching and learning activities. Without the Resident Teachers, the CPI program would not have worked.

These teachers were interested in being Resident Teachers for a variety of reasons. Some teachers wanted to work with other teachers. As some of these teachers indicated through their online journaling, they became involved with CPI because computers were the future, essential to bridging the gap between the theoretical and practical world of learning. These teachers agreed to allow another teacher to come into their classroom for two to three full days. They would observe the one-computer classroom, the two-computer classroom, and the five-computer classroom—each with different strengths and strategies. In the end, the Visiting Teacher took away strategies they could use with their own students given the one computer they had received for completing the CPI training.

For those Visiting Teachers who were paired with the same grade level Resident Teacher, outcomes were the best. There were, however, instances where the Visiting Teacher was paired with a teacher who was teaching two or more grades above or below theirs.

There was training also for the Resident Teachers. During the summer, a one to two day course was developed for the Resident Teachers to learn what to do when someone visited your classroom. They received $3,000 worth of equipment for the classroom or school, a reasonable incentive for participating. The original CPI included one cluster and only three schools. In 2000-2001, all clusters were involved, including 25 new schools, 27 Resident Teachers, and 172 Visiting Teachers completed the course and rotations. In 2001-2002, 12 new schools were added, 15 new Resident Teachers were added, and 100 new Visiting Teachers completed the course and rotations. By the end of the 2002-2003 school year, there were approximately 45 Resident Teachers identified from Kindergarten through high school in many schools across the district.

Resources Needed

It would not be appropriate to describe a program of this magnitude without touching on the resources needed. Initially IBM was generous with support in terms of technology and guidance. Also the principals from the first few schools supported the process by making it possible for teachers to leave their classrooms to gain new knowledge and skills. Releasing teachers for professional development is an expensive proposition for principals, but those who saw the relevance and potential for the program to touch so many children supported without question.

In order for a program like the CPI program to succeed, a school district needs the following resources:

1. A director who is dedicated to the program and fully sees the potential of every aspect of the program. This individual must be willing to support each of the CPI teachers through responding to e-mail and phone calls in a very timely fashion, or visiting their classrooms at any time to give both technical and programmatic support and ideas. Without this key individual, any attempt at this program will most likely fail.
2. A group of classroom teachers who have been through extensive training for using computers within the curriculum, who currently use technology in his/her classroom on a daily basis seamlessly to support the curriculum, and who have mastered classroom management techniques that facilitate uses of computers to enrich classroom teaching and learning activities. These teachers exist in all school districts, but often are hidden in their lone classrooms.
3. A cadre of substitute teachers ready, willing, and able to step into a CPI teacher's classroom and manage the classroom with coherence and continuity following the curriculum set out by the CPI teacher. Much like the program director, the substitute teachers are important components of the program. Without a sufficient number of substitute teachers willing to go into any grade

from Kindergarten through eighth grade, the program will be in jeopardy of not supporting all the teachers.

4. Resources to purchase the computers and software to reward the teachers who complete this program. Giving these teachers the laptop loaners afforded them resources temporarily to gain needed experience and confidence over time. Resources can be achieved through grants or working with corporations to recycle computers into schools. Having incentives for teacher participation is necessary, with the support systems essential to insuring success for all.

Summary

In 2000, Hester Fuller noted that teachers surveyed as part of the National Assessment of Educational Progress (NAEP, 1990) said they wanted more explicit training in strategies to incorporate computers into the classroom with time to actually develop these skills. Reforms do require time to practice to ensure the newly acquired skills are not transient (Cohen, 1987). Similarly, Becker and Ravitz noted in 1999 that there are three resources critical for teachers to institute new practices: "opinion supportive of change, social and informational support that provides practical help to those attempting changes, and a sufficient supply of needed resources that can help demonstrate success to a skeptical outside world" (p. 360). Through the planning efforts of the teachers and the IBM staff who came together to develop what became the Continuous Practice Improvement Model, explicit training opportunities for strategies to incorporate technology into classroom activities were established and time was allowed for these teachers to practice the new skills. In addition, the ongoing support available through the online CPI Web site and the continuous visits by the educational technology specialists helped these teachers to become more and more comfortable with incorporating computers into their lessons and student activities. Support for change was evident, from the principals to the educational technologists to the teachers themselves. This is what CPI was all about – teachers teaching teachers, modeling best practices, all supported by principals who understood what it took to make change happen. What more could you ask for?

As noted in Chapter II, change is very difficult to achieve in general, and appears to be even more difficult in schools. The CPI model was all about change and providing the resources and support necessary for teachers to not be afraid of change. IBM saw this potential back in 1995, and the School District of Philadelphia made it happen through CPI. Due to financial barriers the CPI model was discontinued in Philadelphia in 2003.

References

Becker, H. J. & Ravitz, J. (1999). The influence of computer and Internet use teachers' pedagogical practices and perceptions. [Electronic version]. *Journal of Research on Computing in Education, 31*(4), 356-384.

Coggins, C., Melley, D., & Suh, G. (2000). *IBM project management redefined: The IBM Project management model meets Reinventing Education*. An IBM White Paper. West Chester, PA: IBM.

Cohen , D. K. (1987). Educational technology, policy, and practice. *Educational Evaluation and Policy Analysis, 9*(2), 153-170.

Cromwell, S. (1999). Critical Friends Groups: Catalysts for change. Education World. Last accessed on January 15, 2006 from (http://www.educationworld.com/a_admin/admin/admin136.shtml)

Dockterman, D. A. (1998). *Great teaching in the one-computer classroom.* Fifth edition. Watertown, MA: Tom Snyder Productions.

Fuller, H. L. (2000). First teach their teachers: Technology support and computer use in academic subjects. *Journal of Research on Computing in Education, 32*(4), 511-537.

Glasser, W. (1993). *The quality school teacher*. New York: HarperCollins Publishers, Inc.

Grossman, P., & Wineburg, S. (2000). *What makes teacher community different from a gathering of teachers?* Seattle, WA: University of Washington, Center for the Study of Teaching and Policy. Retrieved on April 12, 2006 from http://depts.washington.edu/ctpmail/ PDFs/Community-GWW-01-2001.pdf

Kanter, R. M. (1999). From spare change to real change: The social sector as Beta site for business innovation. *Harvard Business Review, 77*(3), 122-32, 210.

Litow, S. S. (1999). *Reinventing Education grant program*. Armonk, NY: IBM Corporation.

McLaughlin, R., & Melley, D. (nd). *IBM bridges digital divide through Reinventing Education partnership.* West Chester, PA:IBM.

McLaughlin, R. & Melley, D. (1999). *IBM Reinventing Education grant project completion report*. West Chester, PA: IBM.

Moersch, C. (1995). Levels of Technology Implementation (LoTi): A Framework for measuring classroom technology use. *Learning and Leading with Technology*, November, 40-42. Retrieved on January 15, 2006 from http://www.learning-quest.com/software/LoTiFrameworkNov95.pdf

Moore, D. (n.d.). *CPI information for brochure: Continuous Practice Improvement*. [Brochure]. Philadelphia.

Moore, D. (2001, September). *Effective Lesson Planning*. Power Point Presentation.

National Assessment of Educational Progress (NAEP). (1990). *Profiles of literacy: An assessment of young adults*. Princeton, NJ: Educational Testing Service.

Schifter, C.C. (2001). *Preliminary report from review of the CPI teacher journals*. Unpublished manuscript, Temple University.

Schifter, C. C. (2006, April). *Technology and teaching: Comparing 3 professional development programs*. Paper presented at the annual meeting of the Society for Instructional Technology in Teacher Education (SITE) in Orlando, FL.

School District of Philadelphia. (n.d.). *Children achieving action design document*. Retrieved on August 27, 2004 from http://www.philsch.k12.pa.us/Children_Achieving/exec_summary.html

School District of Philadelphia. (n.d.). *CPI Summer Institute Flyer*. Philadelphia.

Slavin, R. E. (1997). *Educational Psychology: Theory and practice*. Boston: Allyn and Bacon.

Spielvogel, R. (2001). *IBM Reinventing Education: Research summary and perspective*. Retrieved on July 12, 2004 from http://www2.edc.org/cct/publications_report_summary.asp?numPubId=52

Spielvogel, R., Brunner, C., Pasnik, S., Keane, J. T., Friedman, W., Jeffers, L., John, K., & Hermos, J. (2001). *IBM's Reinventing Education grant partnership initiative: Individual site reports*. Retrieved on July 12, 2004 from http://www2.edc.org/CCT/admin/publications/ibmsite.pdf

Tree-Song. (n.d.). *The Medicine Wheel – Circle of Life*. Retrieved on March 30, 2006 from http://www.spiritualnetwork.net/native/medicine_wheel.htm

Vilceanu, M. O., & Schifter, C. C. (2002). *TLCF grant final report: CTEC CPI Project.* Unpublished manuscript, Temple University.

Endnotes

1 It should be pointed out that the Apple Classrooms of Tomorrow (ACOT) studies conducted by Apple Computers occurred during the late 1980s. Apple Computers supported these efforts with extensive technical support, hardware, and software, and studied the results for ten years. See Fisher, Dwyer, and Yocum, 1996 to learn more about the ACOT studies.

2 Clusters were akin to mini-school districts within the overall Philadelphia School District. Typically centered around a high school, the feeder middle and elementary schools would be included in the cluster. Each cluster had a

"superintendent," along with curriculum leaders and technology leaders. The idea was for there to be local support for schools that were connected. There were approximately 22 clusters. This model was abandoned with a change school district leadership in the early twenty-first century.

3 "Wired for Learning" was later developed into "Learning Village" with more resources and tools for the teachers to access.

4 Act 48 requires all certified educators in Pennsylvania to "complete six college credits, six credits of continuing professional education courses, 180 clock hours of continuing professional education courses, or any combination of collegiate studies, continuing professional education courses or learning experiences equivalent to 180 hours every five years." To learn more about Act 48, see the guidelines found at http://www.teaching.state.pa.us/teaching/lib/teaching/PEPGuidelines5-03.pdf. (last accessed January 15, 2005)

5 A school-based Technology Teacher Leader (TTL), designed by Arlene Kramer and Jerry Snyder as noted in Chapter III, was identified as someone in the school, could be the computer teacher or not, who would be a local technology support person for teachers in that school. The original concept of the TTL was a teacher who would have release time from teaching responsibilities to support computers in classrooms. Due to budget constraints, this arrangement was rarely achieved.

6 It is important to note that assignment to either the basic or intermediate cohorts was based on the teachers' responses to the LoTi survey. If the teachers were truthful in their responses, they were assigned to the appropriate group.

7 The location, while centrally located at the Educational Technology Group offices and labs, had limited parking available and no eating establishments near by. When teachers were coming to a location that was inconvenient to locate from 9am to 2pm, having no opportunity to obtain lunch or refreshments added to the inconvenience.

8 The LoTi Framework is based on the extensive research by Dr. Moersch. It is internationally recognized as a tool to "accurately measure authentic classroom technology use." For more information see the LoTi Web site at http://www.loticonnection.com.

9 Most often, Substitute Teachers were retired teachers from the School District of Philadelphia.

10 One interesting suggestion from the program was identifying one or two students in the classroom who already were technology literate. When students were using the computers in the classroom and the teacher was teaching a lesson to a group across the room, if there were a technical glitch for the computer-using students, they were to approach one of the technology student gurus first. The technology student gurus were to be the first line of defense for technology

problems, followed by the teacher if the students could not solve the problems. Most often the problems were simple enough for the students to solve so the teacher could continue with his objective of teaching the lesson at hand.

11 The process changed after feedback from the first CPI cohort, with visitation days reduced to only two days.

12 In the period of time when CPI was available to teachers, the School District of Philadelphia had a set of curriculum frameworks for each grade level that provided guidance to teachers for teaching and learning outcomes but did not dictate curricula. These were replaced in 2003 with standardized curricula for each grade level.

13 "Harder areas" can be interpreted as schools in low income neighborhoods where there were consistent evidence of student behavioral problems, especially in middle schools.

Section II

Teachers Voices:
Early Childhood Teachers (K-2 Grades)

Memories of early childhood education for many people are of finger painting, story telling, learning the alphabet and numbers, and singing songs, among many other memories good and bad. For many teachers the expectation of classroom experiences in these formative years is of experiential learning with lots of hands-on, minds-on activities, but not about the use of computers in supporting these learning goals. CPI introduced teachers from Kindergarten, first and second grade classrooms to how computing can not only support their goals for learning, but also expand the scope of opportunities for students to express how and what they have learned. The next three chapters present the stories of seven kindergarten, six first grade, and six second grade teachers and how participating in the CPI program, as either a CPI Visiting Teacher and/or CPI Resident Teacher changed their approaches to teaching. These stories will be introduced with a description of the school where they teach, followed by a description of how the CPI program impacted their classroom and their teaching. The teacher names are pseudonyms to protect their anonymity. Specifically, they were asked four questions.

1. In what context were you teaching when you participated in the CPI program?
2. What were the personal conditions (feelings, hopes, needs, moral dispositions) that brought you to want to infuse technology into teaching?
3. What factors outside yourself (policies, school, teaching, other teachers) influenced your decision?
4. Considering over time, what has changed in your teaching practices since CPI?

Chapter V

Beyond Finger Painting: Kindergarten

Technology shapes the growing mind. The younger the mind, the more malleable it is. The younger the technology, the more unproven it is (Healy, 1998, p. 17).

Imagine, as best you can, your Kindergarten classroom from way back when, or that of your children's Kindergarten classroom, or perhaps one in which you teach or have taught. These are spacious rooms with tiny tables and chairs made especially for the small children who inhabit them during the school year. These small children are the future and their education in the early years is so important for establishing a set of skills and knowledge base to support their education for years to come. While all teachers are special, Kindergarten teachers take on the role of weaning children away from the home into the milieu of schooling. They often are a surrogate parent for these children who are away from home in a strange place without Mom or Grandmother or any other relative or familiar figure for the first time in their lives.

Making the Kindergarten classroom warm and inviting is so important for these small learners to be comfortable and make the classroom their own. There should be lots of resources available for these children to explore and learn through what is referred to often as hands-on/minds-on activities. Teaching centers, or areas set aside in the classroom for specific small-group teaching and learning activities, are important areas in Kindergarten classrooms, as they are in all early grade classrooms. It is through these teaching centers that children can investigate books, materials and ideas, from science to their environment, their community, and computers. Often Kindergarten classrooms do not have many computers available because there are educators and others who believe small children should be doing hands-on work rather than computer work (Healy, 1998). However, having computers available to expand the repertoire of learning experiences, not as the only learning experience, reflects the view that education of this generation of learners should be from the 21st century perspective rather than that of the mid-20th century.

This chapter presents seven CPI Kindergarten teachers who embraced computers in their classrooms for many different reasons, and were successful for different reasons. They were in different schools with different populations of students and from different areas of Philadelphia. The names of the teachers have been changed to pseudonyms to honor their requests for anonymity, and schools are described in general terms. But their stories are poignant for how the CPI program impacted their teaching and learning environments.

Table 1 gives the reader an overview of the teachers presented in this chapter. The information provided includes how many years of teaching experience they had prior to participating in the CPI program, what grades were included in their school, how they characterized their prior use of computers/technology prior to the CPI program, and whether they were a Visiting Teacher or also asked to be a Resident Teacher. This information may help the reader in deciding which teacher story is close to theirs.

Table 1. Kindergarten teacher summary

Name	Years teaching	Grades in school	Prior computer usage	CPI teacher
Joyce	38 years	K – 8	Some knowledge	Both
Patrice	20+ years	K – 4	Novice user	VT only
Ellen	10+ years	K – 8	Some knowledge	VT only
Marta	10+ years	K – 5	Novice user	VT only
Ronnie	36 years	K – 5	Novice user	Both
Marla	5 years	K – 4	Novice user	Both
Eleanor	20+ years	K – 8	Novice user	VT only

Joyce

When Joyce was interviewed, she had recently retired from the School District of Philadelphia. She had been a teacher at a school in a large urban Kindergarten through eighth grade school in southern Philadelphia where she had taught for 38 years. The school was classified for this book as having a middle level of student enrollments (see the Preface for an explanation of enrollment levels). She started out teaching second grade, but spent most of her career teaching Kindergarten. The majority of students were African-American, with a very small population of English speaking Hispanics and Asians, and an even smaller population of Caucasian children. Approximately 90 percent of the children were from low-income families; thus, all the children could receive the free breakfast and lunch, if they chose. Attendance was between 85 percent and 90 percent in any given year. The turnover in children per year was great at approximately one third, as averaged over the years 2000 through 2005. According to Joyce, the main elementary school building was clean and in good repair, and, with the exception of the lunch room, was usually very orderly (anonymous, personal communication, October 28, 2004). In 2002 this school became a partnership school, managed by a for-profit company rather than the School District of Philadelphia. This was an important outcome, because the curriculum and the way the school was managed changed. Within two years of these changes Joyce was eligible to retire and she did.

Joyce was one of the first CPI Visiting Teachers to be asked to be a Resident Teacher in the CPI program after the program was expanded district wide. She said she was selected because she was identified as being a teacher who was comfortable with classroom management overall and using computers in teaching, open to teachers observing her classroom, and willing to share her experiences with others.[1] Joyce had taken a state subsidized technology program for teachers at a local university back in the late 1970s. That program was not a great success for Joyce, but she came away not afraid of using computers. At one point her classroom was situated in the nearby high school, where she found old computers not being used and some "kid-friendly" software. She shared:

I saw how interested the children were in using them and how great they were for remediation. I started taking [school district] technology classes [again] and was hooked. If I got into trouble I felt there was someone I could go to for help.[2] One of the instructors told me there was nothing I could do that couldn't be undone. That instructor encouraged me to do the CPI program. I wasn't sure that I had enough computer know-how to help other teachers. Then I met the Director. She had such a positive approach to things. There was never a doubt in my mind that I could do the program.

Joyce had five computers in her Kindergarten classroom: four iMacs and one very old Apple computer. They could not access the Internet with the Apple, but it was good for word processing, drawing and some software programs, including a good "mouse" exercise program she found. She had attended the summer CPI program for three days and observed four CPI teachers there—three teachers from the elementary school and one teacher at the middle school. Also, CPI, at that time, afforded the opportunity for Resident Teachers to visit other Resident Teachers' classrooms, from which she reported she continued to gain new ideas for incorporating computers into what she was already doing in her Kindergarten classroom. Joyce further noted:

One of the most enjoyable aspects of CPI was the opportunity to share ideas with other teachers, and for them to share ideas with me. It was a mutual sharing experience for me. I always learned something new from each teacher who visited my classroom.

Over the two years [I was a CPI Resident Teacher] I had about 10 Visiting Teachers in my classroom, most who were either Kindergarten teachers or early grade teachers. I remember one teacher, who later became a CPI Resident Teacher, who was unsure of what Kindergarteners could do on a computer. This teacher was afraid of having children go online without direct supervision by herself. Through her visiting my classroom, she learned how to do exactly that. I always set up heterogeneous groups for working on the computer, with one child with better technology skills as the "captain" of the group. This child would troubleshoot problems, or help the less technology savvy children learn how to access programs or information. I had set up my computers such that the children could access their programs without access to more delicate computer programs and/or unauthorized Web sites. Each child had their own folder where they could keep any unfinished work. The Visiting Teacher learned how to do this as part of the visitation. In the end, this particular teacher changed her practice, and became so adept with infusing computers into her classroom that she was asked to be a Resident Teacher.

Not all teachers who came to visit me were as open to learning. I remember one teacher who basically went through the motions because her primary interest in CPI was to get the computer. This doesn't mean she didn't gain knowledge from the visit, but she was not as engaged in the process as those who were genuinely interested in changing classroom practice and becoming a better teacher overall.

Joyce also mentioned the Learning Village CPI Web site as a good addition to the program. Teachers could communicate with other CPI teachers, express "aha" moments or concerns, and basically feel comfortable that they were in this experience with other teachers like themselves. They could post materials to share as well.

Lastly, Joyce was one of the Resident Teachers who taught the online version of the course during the last year of CPI offerings. Only a few teachers were selected to take the online course since they needed to have basic computing skills well established. She said:

Basically, these teachers were a bit more technology savvy, since they didn't need to start from the beginning with learning how to turn on the computer and which way to hold the mouse to be effective. These were teachers who were able to move at their own pace, some finishing assignments quickly and others asking for more time when the course was basically coming to an end. I believed these teachers were equally prepared for their visitations as those who took the face-to-face course.

Joyce is but one example of a teacher who, like the Philadelphia teachers described in Chapter I, became "hooked" on computers almost from the beginning. Using the technology itself was not the tricky part of incorporating computers into classroom routines, even for the Kindergarteners. Managing a classroom full of eager, busy, curious, and energetic Kindergarteners such that they were productive and mastered skills and knowledge was what set these teachers apart. Joyce was not afraid to try new approaches to teaching, and was eager to share what she had learned with others.

Patrice

Patrice taught Kindergarten at an elementary school located in northern Philadelphia. She participated in the original CPI program from the late-1990s and had been teaching for over 20 years when CPI came along. Her school was a Kindergarten through fourth grade school and was classified as having a middle level of student enrollments. The neighborhood was considered low-income. She described the student population as multicultural, from varied backgrounds, and with 65 percent as non-English speakers. About 50 percent of the students came from Spanish speaking homes, and there were Middle Eastern students, different Asian populations, and African American students. Over 80 percent of the children in this school came from low income families and were eligible for free breakfast and lunch, if they so chose. Unlike the school where Joyce taught, the turnover in the student population was about 17 percent, and attendance was consistently between 91 percent and 92 percent.

In spring 2004, Patrice had eight computers in her classroom, most of them PC-type computers and at least five to six years old. Her classroom was in the basement of the building, but very airy and spacious with a great deal of technology available to her and the children. "Our principal was really supportive of technology in teaching, and encouraged as many of us as were interested to participate in CPI."[3] Patrice rode to school every day with another teacher at this school, and they both participated in the CPI program. Patrice noted (anonymous, personal communication, June 7, 2004):

It was great, because we would talk about different aspects of the program to and from school. Neither one of us was particularly savvy about computers. I would say we were more than a little apprehensive about what this could mean for our classrooms. We encouraged each other to try new things.

She talked about initially having a computer connected to a large screen TV in one open area of the classroom near a large rug used for group activities, and asking the children to gather around the monitor. Together they would read a story, illustrate the story, and explore how the story fit into their lives.

I use the computers for many things, not just for the software we have available. I have many children in my classroom who do not speak English at all. So I use the computers to help them become familiar with the English alphabet, to see their name typed using this alphabet and how this relates to how their name appears in their native language and/or alphabet. I really believe it's been a powerful tool in orienting non-English speaking students into my classroom. They help each other, and get so excited when they are successful and discover new things and talents they didn't know they had.

Patrice's classroom was truly atypical in having eight computers available. They were not new machines, but they all worked. They did not all access the Internet, but she was not concerned about that at all. Patrice learned to use the power of computers for both whole group activities and individual activities, as she described. Student success in her classroom was most important, and she celebrated those successes by hanging work around the room and in the hallway for all to see. Patrice had a very supportive principal who encouraged his teachers to explore new ways of engaging students in learning, one of those essential ingredients noted by Arlene Kramer and Jerry Snyder in Chapter I.

Ellen

Ellen participated in CPI in 2000-2001. She had been teaching at her school for one year, and was teaching seventh grade at that time, but changed to Kindergarten the next year. Prior to starting as a teacher at this school, she had over 10 years of experience with early childcare. She had been a director of a childcare center for at least one year. When Ellen was interviewed she was teaching Kindergarten for the third year and was very happy with her classroom. Ellen taught in a Kindergarten through eighth grade school in the upper regions of northern Philadelphia classified for this book as having a high level of student enrollments. Students were from many different nationalities and ethnicities. Languages spoken included Russian, Arabic, and Spanish, along with many Asian dialects. Approximately 50 percent of the student population was African American, 25 percent were Latino or Hispanic, 18 percent were Asian, and the remaining population was considered Caucasian. The school was not in an economically distressed area, but 70 percent of the students were classified as disadvantaged and eligible for either free breakfast or lunch.[4] As with Patrice's school, the student turnover rate was low at about 15 percent.

At the beginning of the 21st century, the school built an annex where the Kindergarten, first, and second grades were located, as well as the lunch room for the entire school. This building was bright and very well equipped for technology throughout. When asked why she wanted to participate in the CPI program, Ellen said (anonymous, personal communication, December 7, 2004):

I wanted to participate in CPI because I had two computers in my classroom, both iMacs. They were outdated, and I was more comfortable with a PC. My dad is a computer programmer, so I think computing may be in my genes. I've only owned a computer for about 10 years, but it was a PC instead of an Apple. I only started using a computer in college. I wanted to know more how to use spreadsheets and word processing software, how to copy and paste, and other basic computing functions. So CPI seemed like a good fit for me. I got one computer for CPI and took another course and got another computer, both of which follow me at my school. My older computers were taken out of the classroom for another purpose, but I still have the two computers I got through professional development. I use them with the students to learn their letters, to type their names. I use games for eye-hand coordination. The students use Kidpix® for painting and drawing.

CPI gave me a lot of options. Before CPI I felt I could not do much because I was more comfortable with the PC instead of the Apple. I learned how to block certain Web sites so students wouldn't get to inappropriate materials.

With Kindergarten students Ellen used the computer as a teaching center. Each student got a chance to use the computer to work on their letters and typing their own name. She described the children as not fearful to use the computer, and they were learning their letters. The children knew what programs they could use and which they could not. And like Joyce, she had "experts" for the different programs so if she were conducting a guided reading session the students would go to the software experts for help. "I learned this from CPI as well. I do not believe I could achieve what I do now without the [CPI] professional development."

Overall, Ellen participated in CPI to improve her own skills, to get an updated computer for her classroom, and to help her children learn. She stated she achieved her goals. In fact, she was called upon by other teachers to troubleshoot problems. Now that she was computer literate she had the confidence to work with other teachers when they were having problems. She said she might not always know how to help them, but she did understand what they were saying and could refer them to the best person.

One example of how the computers in my classroom have helped a Kindergarten student is one of my students who had difficulty communicating with others. The students had to write a sentence using the writing prompt 'I am thankful for…' and illustrate it with a picture. We were learning about family, and this was intended to be a writing exercise about being thankful for family. This child didn't want to write about his family for a lot of reasons, so he wrote about The Incredibles™ and his picture was a large "I." The students read their sentences aloud to the class and he said it so fast that no one could actually understand him. I worked with him using Google™ to search for the 'I' and cartoons, figuring this was an action figure. Sure enough, when The Incredibles Web site came up, the child said 'That's it, Ms. H. That's it!!!' He was so excited, and I was able to help him with pronunciation. This was something I really could not have done before becoming comfortable using the iMac and Internet searches using Google.

Ellen did not start out to be the computer expert in her school; but along with the other teachers who participated in CPI, she gained the confidence to share her newly gained knowledge and confidence with others. She was not the only CPI teacher from her school. You will meet several more teachers from this school who also became resources for other teachers.

What is also of interest from Ellen's experience was how she moved from thinking her students could not use computers to understanding the power that computers could bring to helping her students express themselves and communicate with others. In the Kindergarten classroom, where students were just learning their letters and how to recognize letters and simple words, the computer was one additional resource for her. Like Patrice, Ellen used the two computers in her classroom to facilitate student learning and personal expression to demonstrate success for all. Using her student "experts," students in Ellen's class could help each other, much in the way the Medicine Wheel legend in Native American lore and in CPI where each member of the community provides valuable knowledge and assistance to other members of the community for the good of all. (See Chapter IV for further description of the

Medicine Wheel legend.) Also, like Patrice, there was evidence of student learning and engagement all around the room, with student art work posted everywhere to celebrate their success and learning using the two computers available there.

Marta

Marta taught at a very small school in a northeastern Philadelphia neighborhood. In comparison to the schools where Joyce, Patrice or Ellen taught, this school had approximately half the number of students, or was classified as a low enrollment school, and was a Kindergarten through fifth grade school. The school was only two floors rather than three or four floors, as are so many elementary schools in Philadelphia, with no need for an annex. Marta indicated the neighborhood was considered blue-collar and very stable (anonymous, personal communication, December 10, 2004.) Most of the people who live in this area were described by Marta as firemen, policemen, and teachers, with a lot of stay-at-home moms. This meant, according to Marta and others at the school, that the Home and School Association was very active and involved with the school. The student population was mixed and different from that of Joyce, Patrice or Ellen, with over 50 percent classified as Caucasian, around 28 percent as African American, and 15 percent as Latino or Hispanic. Only about 45 percent of the students were from low-income families and eligible for free breakfast and lunch. As with Ellen's school, the student turnover rate was about 14 percent, making this a very stable student population and attendance was consistently around 93 percent.

Marta had been teaching at this school for over 10 years when she participated in CPI in fall 2000. She had earned previously a master's degree in elementary education and concentrated in computers from a local college, but she said the master's degree program was mostly about evaluating software not using software with children. "I came away from that degree program thinking [the computer] was not useful for little children. I was comfortable with the Apple, but CPI opened my eyes to the possibilities." Along with several other teachers at this school, she participated to get technology for her classroom and to gain new skills for herself to support her children's learning.

The advantage of taking CPI with others from your school was that you had this group to support you in your successes and failures. The children didn't have computers at home, but took to them readily. It was us [the teachers] who had the problems.

Now I have my students using digital cameras and camcorders to create projects for the school. I actually put together the graduation slideshow with the help of my students. Only because of CPI did I learn how to be comfortable with computers and how to use them with children. I would never have considered any of this before CPI (anonymous, personal communication, December 10, 2004).

Marta was the first Kindergarten teacher who spoke about using digital cameras and camcorders with her children. This will be revisited with upper grade teachers, but Marta was unique in her embracing how Kindergarten children could use such technology successfully. Indeed, her approach, after the CPI program, was that there really was no reason not to ask children to try new activities. As she noted, the children took to the computers and accessories much more easily and readily than the teachers/adults did, so why not explore other options. She did not speak of the students learning their letters and how to type their names, but her children were being very creative in learning how to explore and communicate knowledge through multimedia.

Important about Marta's story was her mentioning that a number of teachers from her school participated in CPI together to support one another through the program and afterward. This was also true for Patrice, indicating how important it was for teachers to have a cadre of colleagues with the same set of technology-related experiences to give them a local support network. Additionally, Marta had a very supportive principal, confirming the findings reported in Chapter I indicating the key role the principal plays in the acceptance and integration of computers into the classroom. While she did not mention other teachers, Ellen also had a core group of eight to 10 CPI teachers and three to four Resident Teachers in her school who were available for discussion of new ideas.

Ronnie

Ronnie taught Kindergarten at a school in northeastern Philadelphia. The year 2005 was her 26th year at her school and 36th year teaching. The school was a Kindergarten through fifth grade school and was classified for this book as a low enrollment school. There were more than 15 languages spoken at this school, and less than half of the students participated in the Title I free lunch program. As with Marta's school, approximately half of the student population was classified as Caucasian, with the remaining student body divided between African American, Asian, and Latino or Hispanic. Ronnie noted that the students consistently scored highly on the state standardized tests, which was not necessarily the case for the other schools in the district. Attendance rates were consistently around 93 percent, and student turnover was 17 percent over an academic year, again a very stable school.

There were 23 teachers at the school, most of whom had over 25 years experience in teaching. As Ronnie put it, "Once a teacher comes to teach at this school, they usually remain till they retire from the school district" (anonymous, personal communication, June 6, 2005). She participated in CPI in fall 2000, and noted:

Before CPI, using computers was a trial and error process for me. I had a PC at home and my son set up a menu program to make it easy for me to get to programs I used. All I had to do was type a letter to get into a specific software application. Then I moved to a Mac, which was a different world all together. I learned programs for children, but otherwise used computers minimally. I did not have a computer in my classroom, and there was no computer teacher at the school. I learned to take my students up to the computer lab to do work on the computers.

I wanted to participate in CPI because I wanted more information about computers for myself. I 'sounded' like I knew more than I did and wanted to fill in the gaps in my knowledge to do more for myself and with the children.

I learned how to integrate computers more into all aspects of teaching and learning. I expanded my use of Kidpix, spreadsheets and more. Now I have six computers in my classroom, all from different sources.

Ronnie was a Visiting Teacher who went on to become one of the Resident Teachers. "I wanted to be an RT because I liked working with other teachers, thought I had something to share about using computers in classrooms, had been a new teacher mentor, and liked the idea of interacting with adults. We would sit and discuss various issues of using computers in classrooms, and then I would help them set up a system for their classroom." Ronnie pointed out the large screen TV and a switch box she received for her classroom in return for her being a Resident Teacher. This setup allowed her to use the TV as a monitor with a computer, a VCR, or a digital camera. She reported that this arrangement made her classroom very versatile. When her classroom was visited in June 2005, the children were watching a movie from the VCR, and reported that they used the computers a lot for various projects.

Ronnie's Kindergarten classroom was divided into several areas. Her six computers were on a computer kiosk that allowed six students to work simultaneously, or for her to use the TV monitor to share a program with the entire class or group of students. As with Patrice and Ellen, student work demonstrating their creativity using the computers was displayed throughout the classroom and in the hallway. Indeed this school's teachers appeared to celebrate student achievements by sharing these works with other students and teachers throughout the school's hallways. Again, a number of teachers from this school participated in the CPI program, but Ronnie was the only Kindergarten teacher who did. She came into the CPI program as someone eager to learn to make these machines work for her students better than through her "trial and error" approach. In the end, not only did she accomplish her goals, but she was able to share her new knowledge with other teachers across the district at a Resident Teacher. It is interesting to note that, like Marta, Ronnie also used digital cameras with her Kindergarten students.

Marla

Marla participated in CPI in fall 2000. She had been teaching for only a few years in a Kindergarten classroom in an urban elementary school in a southern Philadelphia multi-ethnic neighborhood. The school contained Kindergarten to fourth grades with a middle level of student enrollments. The school population was about 75 percent African American, and 24 percent Asian. Between the years 1999 and 2005, the student population dropped by 40 percent, probably due to changes in neighborhood demographics and development of charter schools in the district. Over the years, other than the drop in enrollment, student turnover during the school year was approximately 20 percent.

It was important to note that Marla started teaching Kindergarten at this school but became the computer teacher in 2002. Like Jackie and Ronnie, she was first a Visiting Teacher learning from the CPI program, and then she was selected to be a Resident Teacher because of how well she integrated technology into her classroom management strategies and student learning activities. Marla put it this way (anonymous, personal communication, October 8, 2004):

In 1981 I went to a business high school where they had computers. Then I went to a regular high school where they didn't have computers. That was in the early to mid 80s. After graduation I got a job and was afraid of computers. My boss sent me for a computer class, which I took, and then basically only used word processing. I took the class and felt better and used the computer for work.

In college I was exposed again to computers and to e-mail, CD-ROMS, software, games, and such. I was doing e-mail way back with the old black and white monitors and the orange blinking cursor. My instructor wanted us to e-mail him and each other and get our assignments that way. The e-mail addresses were long, long, but that was back in the day. We got our assignments that way.

As a teacher, Marla saw where computers were heading. At her first teaching position school, she had a technology class and was both impressed and intimidated. She said she would go to the class and learn different skills, but she still did not feel comfortable, again noting that she felt intimidated. She seemed to have a hard time "figuring out" computers.

There was a teacher in the school district who was a major influence on me. She was in my region and I do believe she was a motivating factor for me. I wanted to start a PC club at my school. This teacher recommended another teacher at [my current school] who was just fabulous. [The second teacher] suggested that I transfer to this school, and I did. So, she did projects [like] a pet census and a fruit book that went on the Internet. Both of these teachers encouraged me to take computer classes. I love learning new things. I had two Apple IIEs with "ease" buttons in my classroom at that time, but wanted more.

I took the CPI course and one other course and got two iMacs for my classroom. I saw endless possibilities for the classroom, especially for lessons. I wanted to be up with the times. Some people were interested in the latest fashions, I wanted the latest technology. I learned from my Resident Teacher how to manage the students and different programs, how she used the computer for her lessons, how she rotated students to use the computers, how she imported graphics into documents, her Web site, and overall how computers impacted her classroom. She created a news letter and a Web site. PCs are multimodal for the kids. They see it, hear it and work with it. It's great!

I didn't have a computer at home. I remember with that laptop I could not get into the Web for long at home because we had pulse dialing at home. So I worked mostly at school. Now we have [broadband] at home and I'm on every five minutes getting lessons and information. With all the sites, you can supplement your lessons a lot. Back then [in 2000], if you were doing a lesson and had to run off something, you could go to the computer and print it out and run it off at lunch time. It's a portal for information. I would never have done any of this without CPI.

At the start of CPI, it was kind of like everyone did their own thing. Then we started sharing, and with the [CPI] Web site we can share. I'm someone who likes to tell people what I just learned, whether you need it or not. Some of the teachers still send resources to other CPI teachers. You go to different Web sites and get ideas and share with other teachers. That's what it was all about, camaraderie. If they brought CPI back in the district, I would go back to learn new things – Continuous Practice Improvement.

CPI was phenomenal. You got a laptop, a computer and a printer. You got time to work. They, the CPI instructors, set the tone which was 'you can do this.' Some teachers were told by their principal to take the course. They were not as comfortable as some of us who chose to be there were. The announcement said you had to have some knowledge of computers, but some of the teachers had minimal knowledge. The people teaching CPI had the patience to work with us. We put up our Web site, everything was detailed. I think that was where I learned the basics. The curriculum was there.

I would say CPI took my teaching to another level. It gave me additional resources. Back then, you basically talked to the class and had a television monitor for dem-

onstrations. In Kindergarten, you have the different center stations and the children rotate through them giving them quality time with the computer station. This year I see students who were in my Kindergarten class two years ago. My students appear to have more computer skills than students from other Kindergarten classes, probably because they got keyboarding practice and mousing skills working on the computer in Kindergarten.

Marla's Kindergarten classroom was in a back corner of her school and was a long, dark, and narrow room. At one end near the windows in the front of the room was her computer center with four computers. Because she had a number of children in her classroom who were English language learners, she had several software programs available for them in Spanish and a few other languages where, for instance, they could toggle between Spanish and English. She believed this helped the children learn the English version of what they were saying in Spanish, thus helping them learn English. Again, student accomplishments were displayed around the room demonstrating to all the children that they were each successful in their own way. Marla took away from the CPI program that all of her students could learn, they just did not have to learn at the same rate and be at the same place in their learning, just like the CPI teachers did not learn at the same rate or complete the program at the same level of expertise with computers. She used the computers to facilitate this for all her students, not only second language learners.

Eleanor

Eleanor taught in the northwest area of the city at a Kindergarten through eighth grade school with a low level of student enrollments. The neighborhood was described by Eleanor (anonymous, personal communication, May 22, 2006) as well-to-do, with about one-half of the residents as senior citizens who had lived there all their adult lives and one-half as up-and-coming, well educated young adults with children. Many of the children in the neighborhood attended private school, or they came to this school for the early grades in preparation for applying to one of the more academically oriented magnate schools in the district. In late 2006 about 80 percent of the student population was bussed into the school from other parts of the city. Even so, Eleanor reported that the parents of these bussed-in children went out of their way to get their students into this school and were active in the Home and School Association, especially parents of the younger children.

Students are multicultural in that there are those who are Asian, Indian, Brazilian, Venezuelan, among other nationalities. There is a growing English as a Second Language program. The school is introducing an International Baccalaureate program hoping to entice some of the students who choose to leave to stay beyond fourth grade.

Her Kindergarten classroom had a little kitchen and many other resources for her children. Eleanor noted that the curriculum required investing extensive time in reading and mathematics, rather than exploration of new ideas and experiences. But that had not kept Eleanor from bringing exciting adventures into her classroom through the computer she earned from the CPI program, as will be described below. Her Kindergarteners attended computer class for 45 minutes each week where they played math games, learned to use Kidpix creative software to make Mother's Day cards and such, and to work on reading as well. In her classroom, Eleanor had 23 children with no aide, so she had her hands full every day.

I wanted to participate to get a new computer and printer for my classroom. I had a really old computer that students could use to type up journal entries to post, but no printer. No printer made it difficult to post materials around the room, and made it difficult for the students to see the relevance of the activities. My classroom had no Internet access; and even if it did, the computer didn't have enough memory.

After CPI, I had a schedule made up for the students to use the math and reading games I received, make cards for parents, and type out a journal entry or a story they had written. The schedule made it easy for each child to get a chance to explore the "new" tool, but the printer really made the difference in the classroom. The school did not have printers in individual classrooms. If you wanted to print something, you had to send it to the printers in the computer lab, which meant someone from the classroom had to go there to pick up the work. This process meant students who ran the errands missed instruction or opportunities to work with classmates on projects. Having a printer in the classroom made a huge difference in students seeing immediate rewards for their efforts. I used the printer to hang student work around the classroom or to make books for each child.

Another really important bonus from the CPI program was the new computer that was Internet ready. Now Eleanor could bring the world of the Internet and World Wide Web into her classroom for her students. One example she noted was the class watching a nest of falcons on top of the Kodak® Building in Rochester, New York, which she had done annually with her students for a few years at the time she was interviewed. Five cameras were positioned to watch the baby birds hatch and grow over time. They even had a night camera so students could see what was happening over night, if they chose to watch. Eleanor said she would access that Web site at the beginning of class and leave the image on the screen for the children to occasionally observe. The Web site image refreshed every few minutes, meaning the children did not actually see movement, but they could see the development over time.

While CPI brought useful technology into Eleanor's classroom, she did not change her practice as much as other teachers. She also was the only Kindergarten teacher who mentioned the demands of the curriculum. As noted in Chapter II, change is a constant in schools, including changes in the curriculum, new testing schedules, and more. While Kindergarteners are not tested as a result of the No Child Left Behind Act (Department of Education, 2001), Eleanor clearly was pressured to ensure her students met the expectations of the curriculum, leaving little to no time for exploration of how technology could enhance and embellish lessons. This outcome was not Eleanor's fault, as much as it was predicated on the fact that the curriculum was developed without consideration of how technology and computers could support the students meeting the standards and expectations.

Summary

Each of these seven teachers came into CPI with a different set of skills and goals to be achieved through the program, and across these stories are some interesting concepts. Before participating in the CPI program, five out of the seven described themselves as either novice users or basically intimidated by computers in general. The third teacher, Ellen, knew how to use a computer for basic personal use, but had no idea how to make a computer be part of her classroom in meaningful ways. All in all, they each "believed" that computers could help their students become better students and learn better and faster. Dwayne Huebner (1999), in his essay "Politics and the curriculum" suggests that when teachers express that instructional technology, whether that be radio, film, television, or computers, is the future, they are saying so because they have been influenced by the current politics surrounding instructional technology, that more is better. More technology in schools means students will learn better, and to not use instructional technology labels the teacher as out of touch with the modern world, or a Luddite. In this piece, Huebner was not supporting the position that instructional technologies are of use in classrooms, but noting that there are politics surrounding them, expecting them to be available and used in classrooms. Unfortunately, many curricula are written without the infusion of computers, or any technology, in mind; therefore, technology becomes an add-on that may be expendable, or considered such. What was different about these teachers, and those you will meet in future chapters, was that they turned the technology around to support what they were doing, not just because it was politically correct, but because they saw how their students responded. This was all about the students learning and not about technology.

Analysis

As noted in Chapter II, change does not happen easily. The more motivated teachers are to change their practice, the more likely it will happen and be sustained over time. The more viable and useful the "innovation," the more likely the teachers will embrace, learn and use it. These interviews occurred between four and seven years after these teachers participated in the CPI program. The literature on professional development and impact on classroom practice suggests that institutionalized change is the highest level of change (Fullan and Pomfret, 1977) or adoption of a change (Hord, Rutherford, Huling-Austin, & Hall, 1987), and not easily attained through most professional development programs.

Starting with Rogers theory of diffusion of innovations (1962, 1972, 1983, 1995, 2003), each of these seven Kindergarten teachers had worked through the stages of adoption. They did not get stuck along the way, but fully embraced the notion that computers could assist them in their quest to engage their students in learning activities, whether low level (e.g., learning letters) or higher level (e.g., using camcorders to create materials) to even higher levels of asking questions and predicting (e.g., watching Peregrine falcons grow up). The CPI program itself did not teach them how to do these things, per se, but gave them the confidence that all children could learn how to use computers at a comfort level unique to each child—and that this was okay.[5]

Using the lens of the Concerns-Based Adoption Model (Hord, et al., 1987), these teachers had moved beyond concern about how the computers affected themselves or the amount of time it took to manage the students' use of the computers, to how the use of these computers affected the students learning, working with each other, developing responsibilities (e.g., student technology experts), and working with other teachers. Three of these teachers were chosen to be Resident Teachers to share their new knowledge and skills with other teachers. They had moved to the level of integration in their daily activities beyond mere routine use by themselves.

Healy (1998) argued that perhaps it is not appropriate for Kindergarten children to be required to use computers. She suggested that, even using the developmental theories of Jean Piaget, Rudolf Steiner, and Maria Montessori, one might question the use of computers with six year olds who might not be developmentally ready for computerized or digital images. "*Children need time and plenty of practice to 'play around' (often literally) with different types of learning* [italics original]; while we can enrich, expand, and possibly pull a child up a step or two, there is little to be gained (except, possibly, an unpleasant fall) by trying to propel someone to the top steps before the lower ones have strong foundations" (Healy, 1998, p. 135). Her comments resonate with how Cuban (1986) characterized Papert's (1980) approach to LOGO, that Papert's theory of using LOGO with young children bypassed Piaget's necessary concrete-operational stage for ages 7 through 11, and expected the children to be able to function (at ages much earlier than 11) at the formal-operations stages, which Piaget noted began typically at age 11 (Huitt & Hummel, 2003).

So given the wealth of concern about using computers with Kindergarten students, and there is quite a bit of skepticism in the literature, why should these Kindergarten teachers be celebrated? Basically, they did embrace computers for their students, but not as the cure for all problems. The use of computers in these classrooms was but one more way to engage the students in learning many different concepts and skills, for example, their letters, illustrating stories, and more. The presence and use of computers in these classrooms did not eliminate from their practices the use of hands-on/minds-on activities such as drawing pictures with crayons, working in cooperative groups on science projects, or reading books and stories. No, the computer activities were supportive of all activities, and the results were that the teachers might not be as teacher-centered/oriented in their approach to teaching as before, or at all times. And considering "power in the classroom" (Huebner, 1999), these teachers chose to use their "power" or personal capital by opting to participate in the CPI program, to set goals for themselves to become more computer literate, and to bring new resources to bear on their classrooms and students. While there were others who influenced their decisions along the way, as noted by Marla and Joyce, in the end the decision to embrace technology in their classrooms was theirs. They had to take charge of their futures and their classrooms by committing to changing their practice. This will be important for future discussions in chapters to come.

Also in Chapter III, Berman and McLaughlin (1975); Fullan and Pomfret (1977); Sharan and Hertz-Lazarowitz (1980); Little (1989); Ball (1996); Wilson and Berne (1999); Guskey (2000); and Joyce and Showers (2002), suggested that peer coaching or mentoring across and among colleagues in one school, or across schools, could be quite powerful in supporting change in teacher practices. Among these seven teachers, four had cohorts of teachers participating either simultaneously with them or at a time close to their participation in CPI, and available to discuss ideas related to CPI and technology inclusion in their classrooms, and/or to share ideas and examples. These external, ongoing, formal and informal support networks were important in helping these teachers sustain change in practice. CPI fostered these networks through the course and systems put in place.

References

Ball, D. L. (1996). Teacher learning and the mathematics reforms: What we think we know and what we need to learn. *Phi Delta Kappan, 77*(7), 500-508.

Berman, P., & McLaughlin, M. W. (1978). *Federal programs supporting educational change, Volume VIII: Implementing and sustaining innovations*. Santa Monica, CA: Rand.

Cuban, L. (1986). *Teachers and machines: The classroom user of technology since 1920*. New York: Teachers College Press.

Fullan, M., & Pomfret, A. (1977). Research on curriculum and instruction implementation. *Review of Educational Research, 47*(1), 335-397.

Guskey, T. R. (2000). *Evaluating professional development*. Thousand Oaks, CA: Corwin Press.

Healy, J. M. (1998). *Failure to connect: How computers affect our children's minds – for better or worse*. New York: Simon & Schuster.

Hord, S. M., Rutherford, W. L., Huling-Autsin, L., & Hall, G. E. (1987). *Taking charge of change*. Alexandria, VA: Association for Supervision and Curriculum

Huebner, D. E. (1999). Politics and the curriculum. In Hillis, V. (Ed.) (1999). *The lure of the transcendent: Collected essays by Dwayne E. Huebner*. Mahwah, NJ: Lawrence Erlbaum, pp. 15-22. Originally published in A. Passow (Ed.) (1962), *Curriculum crossroads*. New York: Teachers College Press, pp. 87-95.

Huitt, W., & Hummel, J. (2003). Piaget's theory of cognitive development. *Educational psychology interactive*. Valdosta, GA: Valdosta State University. Retrieved on May 29, 2006 from http://chiron.valdosta.edu/whuitt/cogsys/piaget.html

Joyce, B., & Showers, B. (2002). *Student achievement through staff development*. Third edition. Alexandria, VA: ASCD.

Little, J. W. (1989). District policy choices and teachers professional development opportunities. *Educational Evaluation and Policy Analysis, 11*(2), 165-179.

Papert, S. (1980). *Mindstorms: Children, computers and powerful ideas*. New York: Basic Books.

Rogers, E. M. (1962, 1972, 1983, 1995, 2003). *Diffusion of innovations*. Glencoe, IL: The Free Press.

Sharan, S., & Hertz-Lazarowitz, R. (1980). A group investigation method of cooperative learning in the classroom. In S. Sharon, P. Hare, C. Webb, R. Hertz-Lawarowitz (Eds), *Cooperation in education*, (pp. 14-46). Provo, UT: Brigham Young University Press.

U.S. Department of Education. (2001). *No Child Left Behind Act*. Washington, D.C.

Wilson, S. M., & Berne, J. (1999). Teacher learning and the acquisition of professional knowledge: An examination of research on contemporary professional development. In A. Iran-Nejad & P.D. Pearson (Eds), *Review of Research in Education*, Volume 24, pp. 173-209.

Endnotes

1 Remember from Chapter IV that Ms. Maguire selected additional Resident Teachers because they were willing to share their experiences with others and had good communication skills. Joyce is the first example of a teacher chosen to be a Resident Teacher for these reasons in addition to her use of technology in her classroom.

2 One of the local teachers supporting other teachers with technology was Ms. Cook, who was described in Chapter I.

3 The principal at Patrice's school had been the principal at this school for nearly 20 years.

4 In the School District of Philadelphia many students are bussed across town to attend schools not in their neighborhood, which could account for the large number of children from low-income families in a school that is not considered to be in a low-income neighborhood. In addition, Philadelphia has a number of Catholic neighborhood schools, and a growing number of alternative charter schools.

5 It was perhaps also helpful that there was no technology curriculum required by the school district at that time. The district did implement a standardized Kindergarten through high school technology curriculum, based on the ISTE National Educational Technology Standards for Students, in fall 2006.

Chapter VI

Finger Painting to Digital Painting: First Grade

An early fear was that computers would replace other early childhood activities. Research shows that computer activities yield the best results when coupled with suitable off-computer activities (Clements, 1998, p. 6).

Like Kindergarten, the first grade classroom is a welcoming site with small tables and chairs for the equally small children who are typical first graders. For children who did not attend Kindergarten, for whatever reason[1], first grade is when they learn about schooling and the rules that go along with being in school. It is a time to explore ideas, be creative, make friends, learn to be collaborative with others, learn to read better, and much more. While the curriculum is similar to Kindergarten, the level of instruction and expectations for student engagement is slowly becoming higher, but only slightly at this point. Also like Kindergarten, computers tend to be used often for students to express their understanding of a story or concept through creative software. The concept of multiple intelligences (Gardner, 1999), active learning (Dewey, 1938), and applying learning to related knowledge (Vygotsky, 1978) are exemplified in all primary grades, and especially in first grade.

As with Kindergarten classrooms, there has been controversy in the literature about whether computers have a place in first grade classrooms. Barnes and Hill (1983) suggested that computers might interfere with children's needed experiences with real events and objects, limiting engagement and physical activity. In the same year, Karoff noted that computers might help children's memory patterns and sequencing as a tool for thinking and problem solving. Over the years, the fears have been proven wrong for children in early grades, but there are still people who strongly believe that children should be experiencing "real" activities rather than digital ones (Healy, 1998). Whether these concerns are valid or not has not been proven one way or the other, and may take a few decades of students using computers in truly meaningful ways, not just as a reward for good behavior, to know if inclusion of computers in early grade classroom lessons and activities actually makes a difference in student learning and long term outcomes.

This chapter presents six first grade teachers who were not unlike the Kindergarten teachers as far as their previous use of technology. Most of the teachers who came into the CPI program classified themselves as novice to non-users of computers, but were keenly aware that computers were considered by many in the education world (and perhaps even more so in the non-education and political world) as very important to the future of today's students. Students who were not techno-savvy were to be disadvantaged when they moved into the work world, one of the concerns in the discussions of the "digital divide." Students from the "have" schools would be privileged, and teachers by nature want to make sure their students are not disadvantaged when they move on in life. These teachers wanted their students to have access to tools that might provide an advantage.

Table 2 gives the reader an overview of the teachers presented in this chapter. The information provided includes how many years of teaching experience they had prior to participating in the CPI program, what grades were included in their school, how they characterized their prior use of computers/technology prior to the CPI program, and whether they were a Visiting Teacher or also asked to be a Resident Teacher. This information may help the reader in deciding which teacher story is close to theirs.

Table 2. First grade teacher summary

Name	Years teaching	Grades in school	Prior computer usage	CPI teacher
Mary Jo	20+ years	K – 4	Novice user	VT only
Rhoda	20+ years	K – 8	Novice user	VT only
Joan	15+ years	K – 8	Novice user	VT only
Parker	9 years	K – 6	Very comfortable	VT only
Alta	20 years	K – 6	Novice user	VT only
Melanie	30+ years	K – 5	Novice user	VT only

Mary Jo

Mary Jo participated with the second CPI cohort of teachers before the program was expanded across the district. She taught first grade then and now at the same school in northern Philadelphia. In 1998 she had been teaching for more than 20 years. Her school was located in an economically depressed area. She described the students as multicultural, from varied backgrounds, and with 65 percent whose first language was not English. She reported about 50 percent of the students came from Spanish speaking homes. In addition she noted there were a large Arabic population, different Asian populations, and African American students. The school district statistics for this school indicated that indeed nearly 60 percent of the students were of Latino heritage, about 25 percent were African American, and the remaining students were divided between Caucasian and Asian. Over 80 percent of the students were from low-income families making them eligible for free breakfast or lunch, if they chose to participate. Approximately 18 percent of students turned over in any given year, but attendance was at 92 percent on average. For the purposes of this book, enrollments were classified as at the middle level.

In 1998 she had about 30 students in her first grade classroom and she was the only teacher in the classroom. She had a couple of computers but basically did not know how to use them with her students. Mary Jo put it this way (anonymous, personal communication, November 12, 2004):

In 1998, I may have had two computers. They were of great use to me. I did find it difficult to use them with the children until I went through the CPI program. CPI gave me time to actually look at and work through the software available on the computers. At the time the Math [curriculum] series we were using had software that supported the activities, but I didn't have time at school to work through them. CPI allowed me to have time to actually look at the software, watch children using the games. I asked the children in the Resident Teacher's classroom I visited to show me the games and what they were doing, and they were very successful and wanted to teach me the software.

I didn't originally participate in the program because they said the emphasis was on English as a Second Language (ESL) students. The ESL teacher would be coming into my classroom and I thought that would be too much disruption for the children. But that never actually materialized, so I signed up for the second year. I was interested in computers and didn't have any software that was valuable to the children. I didn't get anything tangible for participation, but I did get time to work with the software, see children successfully using the software and games, and got a chance to see another teacher manage the computers as well as her instruction. This latter was difficult for me at that time. So much of the software required reading, but the children were not able to be independent with the software. Software has changed a lot since that time, as have Web sites. I wanted to be able to allow my children to be independent on the computer, and I gained insight on how to do that.

At the school level, Mary Jo reported that the principal was very supportive for all teachers to participate. Mary Jo also rode to work each day with a teacher who was in the first cohort of teachers to participate in CPI.

We were always interested in new things for the classroom. I was not afraid of the computer, per se. I was a Macintosh user, but I changed to IBM through CPI. I had one Mac computer in my room at that time on my desk, but it was not connected to the Internet. I used it for grades, but not instruction. I now have four computers in my classroom, three that work.

As far as my classroom practice, I was able to enable the children to get on the computer on an ongoing basis. Because I was able to see how another teacher could do this, I felt empowered to do the same in my classroom. This was mostly the software that went along with the math curriculum, games and other activities to help them hone their skills. I set the computers up as a center where the children rotated on and off the computers while other children were doing other related activities.

Not long after I completed the CPI program, they changed the math curriculum so the software I was using was no longer relevant. With these changes, the children now use the computers more than before. They have software that supports DIBELS®[2] testing and that relates to how children hear sounds, interpret those sounds, and remember. They use this software 20 minutes a day, so they rotate on and off the computer to accomplish this. The students can find their name and go on. Otherwise, they can type things up, but I don't have them illustrate sentences because that takes too long.

Overall, CPI was good. I enjoyed it for a number of reasons. First, there are few opportunities for teachers to observe a master teacher at work. Second, this was a time when there were consecutive days in the classroom, where it's impossible for the situations to be "staged." In three consecutive days, you saw the problems, you heard the noise level. Third, it was the only time in professional development I felt I was treated like a professional. That came from IBM and their professional development approach. You learned so much by being in a master teacher's classroom—watching one activity, while hearing other activities.

Teachers in the school who participated in CPI would share with other teachers during grade group meetings. They would share software that supported this aspect of the curriculum, or Web sites. One teacher had a large screen monitor in her classroom hooked up to one of the computers. So when she wanted to do whole group instruction about software, the children could watch a bigger screen. I learned a lot.

Mary Jo's experience with CPI was a bit different from other first grade teachers. Like Patrice's experiences from Chapter V, she was a teacher in the first elementary school provided the CPI professional development opportunity. She had access to the original CPI Resident Teachers as well as the first of the Visiting Teachers. Having a very supportive principal who recognized the potential of computers to support student learning was extremely important for Mary Jo, and again supports the theory proposed by Arlene Kramer and Jerry Snyder described in Chapter I.

As noted, the mathematics curriculum changed shortly after she completed the CPI program, meaning the software she had been using was no longer relevant to the new curriculum. Ironically, rather than reporting that she had abandoned student use of the computer, Mary Jo said her students used the computers more than before. Being able to draw upon the experiences of those who had mastered the process of incorporating micro computing into classroom activities easily gave Mary Jo the confidence to explore options for her students' productive use of computers. Having access to other CPI teachers afforded Mary Jo access to resources long after CPI was over.

Rhoda

Rhoda taught at a school located in the northwestern part of Philadelphia. This school was located in a middle class neighborhood where most of the students attending this school were bussed in from other areas of the district. According to Rhoda, most of the neighborhood children attended private or parochial schools rather than the public schools (anonymous, personal communication, June 3, 2005). This school's student population was over 96 percent African American, with very few students who were Caucasian or Latino, and no Asian students at all. Sixty percent of these students were from families classified as low income, making them eligible for free breakfast and lunch. This school was listed as having no English as a Second Language (ESL) students, unlike Mary Jo's school that had a large ESL population. Enrollment at this school had dropped by 20 percent between 1999 and 2005 to a middle level of enrollments as defined for this book, but so had enrollment in the district as a whole. Student attendance maintained an average of 93 percent between 2000 and 2005, and approximately only 17 percent of the students changed schools in a given year. Rhoda taught first grade and had taught at this school for eight or nine years in 2005, and 20 years altogether in the district.

Rhoda did not have any computers in her classroom in 2001 when she signed up for the CPI program. In 2005 she had three computers.

So participating in CPI was a way for me to get a computer for my classroom. In addition, I learned about having the children write journals on the computer, do other writing [on the computer]. The children really enjoyed using computers, which helps them learn. When I took CPI, I had minimal skills in using computers. I would describe myself as a "novice" user at that time. I came away knowing how a computer works, how to use software with my children, and more.

I am still learning, but feel very comfortable using computers. I do reports, enter grades and DIBELS data. Overall, I went from feeling computer challenged to being comfortable using computers with my children thanks to the CPI experiences.

Rhoda's story is not extensive. There were seven teachers at her school who had participated in CPI within a year of one another. Over time they were not equally using the knowledge and skills they gained through the program. One major problem for this school, in particular, was that the building and infrastructure were quite old. Electricity was often the most challenging commodity in older school buildings, especially those built before the 1960s. Since these schools were not built to support a technology-rich classroom, as the school annex options described by Ellen in Chapter V, Melanie later in this chapter, or Jo and Gina in Chapter VII, obtaining the capacity for multiple computers to be used at once was quite a challenge. Using what computers one had in a classroom afforded a technology center for students to use the computer to explore topics, much as they do in the science center or art center. This constructivist approach to students using computers to explore topics follows the theories of John Dewey (1938) and Jean Piaget (1972, 1990).

Joan

Joan taught first grade at the same school as Rhoda. She participated in CPI in spring 2001. She was asked by the school district's local cluster[3] educational technology support person if she would like to participate. When asked about her computer use before CPI, Joan said she was a novice user like Rhoda (anonymous, personal communication, June 3, 2005). Basically, she did not use computers except as a "point and click" operation, meaning she rarely used it any more than necessary. Prior to participating in CPI, Joan indicated that she had a great fear of technology and understood little when it came to computers. She had been interested in using computers in her classroom but was unsure how. Her goals for participating in CPI were to overcome her own fears by increasing her knowledge about technology, and to become a successful computer user before attempting to teach her students to use the computer through her lessons. Joan put it this way:

Using computers was going to be required in the future, so I wanted to learn before I was required to do so. I was afraid of computers and didn't want to be. It took a few sessions to get comfortable, but I got there. I am most comfortable with e-mail.

CPI gave me confidence, made computers understandable. One thing I learned was how to set up files on a computer for the students. Each child would have a folder on the computer for their journal entries, projects, and such. This was a new concept [for me] and seemed to make sense. I learned a lot through CPI, but didn't practice as much as I should have. I now feel like I need a refresher.

I do have a problem in my classroom. While I have three computers, they can't all work at the same time. In fact, when one computer is being used, if another computer is turned on, the first computer freezes. It's very frustrating. The building is due for computer fiber upgrade over the summer, but this does not include electrical wiring upgrades which could be the problem.

After CPI, Joan noted that her knowledge about technology had increased. In particular she had learned how to use slide shows in lessons to captivate her students' attention. She developed a Web page for her classroom, something she never would have dreamed of doing before CPI. Another example of how her teaching was affected by CPI is the lesson Joan developed on graphing by using a spreadsheet program. She was able to help the students understand graphing through seeing the graphs change as she changed the numbers. For instance, if you put the numbers "1, 2, and 3" in a row or column and highlighted those numbers, you could graph them in a pie chart. Students could see the proportion between the numbers helping visual learners (Gardner, 1999) understand the relationships of the numbers to one another. She believed the students were more motivated to learn their numbers because of her ingenuity. Again, Joan would never have even thought of using a computer program to teach mathematics if she had not been personally empowered by the CPI experiences.

Like many CPI Visiting Teachers, Joan established a "computer center" in her classroom. She rotated her students through the center by the use of a checklist, making sure that at least eight students used the computers each day. She noted that, in her opinion, she believed that the success of the center could be attributed to the software that she received through CPI, which were all grade appropriate and held the attention of the students as they learned. She also suggested that there had been a decrease in disciplinary problems since the computers came into the classroom. While she did use access to the computers as a reward, this was not the only reason for students to use the computers since every student had access to a computer each week for learning activities.

In the end, Joan's classroom practices changed because of CPI. Before participating she was afraid of technology and avoided any opportunities even to learn. Encouragement from the regional technology staff gave her the courage to try the program, and the program was all she needed to move from a non-user to one who created a Web page for her classroom.

Parker

Parker taught at an elementary school in the northwestern region of the school district. The school had grades Kindergarten through sixth grade and an enrollment level classified for this book as low. Parker had taught at this school for eight years and in the district for nine years. The school had predominantly African American students with only two percent listed as Latino and one percent as Caucasian in 2006. Attendance averaged 90 percent between 1999 and 2005. Eighty-nine percent of the students came from low income families making them eligible for free breakfast and lunch programs. Over the last seven years, student mobility was at a 20 percent level on average.

At the time Parker participated in CPI in 2001, the educational technologist (ET) in the cluster and the school's computer science teacher had recruited him to participate in CPI. "I always used the older Apple computers I had access to, but the computer science teacher knew I would get a new computer for my classroom through the program. So he and cluster ET recommended that I participate in the CPI program, and then I could do much more interesting things with technology with my first grade students," Parker said (anonymous, personal communication, May 1, 2006). In the end, Parker had four computers in his classroom, three with Mac OS8.6, and one iMac from the CPI program. The additional and newer technology afforded him to establish different ways to engage the students in learning with technology. He noted further:

What I remember most from the course was the training on the Apple iBook loaner laptop so I could practice the skills and do the homework easily either at school or at home, and creating the Web page. Being able to practice what I was learning and creating the Web page were most valuable to me, and I've taken that into the next phase of my career.

At that time in the late 20th and early 21st centuries, Balanced Literacy[4] was the literacy emphasis in the early grades in the district. Parker had "message time" where students would be on the "discussion" rug and he put readings into context for the students. He had "interest centers" for the children, including a technology center. He made sure the software titles were appropriate for them to be able to work independently. He would be there for support, but he wanted the children to develop their critical thinking and technology skills. While he was conducting guided reading groups, other students would be doing independent work—some of which could be in the computer center.

I liked to do digital photography and having pictures of parents when they came to work in the classroom, or just pictures of the children working, or even classroom materials that could be commented on. I would import the pictures into [a word processing document], and ask the students to create text to describe the pictures. I would do the images, and the students did the text. In the end, the students were proud of their own work and their pictures. In addition, I could take these pages and create a book for parents to see when they visited the room, and the book then became part of the reading center for the rest of the year. At the end of the year, the pictures were given to the students to take home.

Parker had a series of CD ROMs with School House Rock® videos. He could use the 15" screen from one of the computers to show the videos to the children to trigger discussion. The classroom management techniques from CPI and his Resident Teacher gave him the skills to have the children do multiple activities at the same time and to not be afraid of the resulting "chaos."

Parker, like Marla in Chapter V, became the computer teacher at his school. When the computer teacher left in 2003, a replacement teacher was hired. The replacement teacher left after one year and the principal asked Parker if he would take on the challenge. Two years later he was still in the position and thriving. A helpful implementation in 2006 was that the district now had a computer curriculum to be required of all schools starting in September 2006. For Parker, this curriculum gave him a context and set of experiences by grade level that were meaningful for all students. While it could be argued that Parker already had the skills for incorporating computers into his classroom, he moved his level of use to maximize the resources he had in his classroom.

Alta

Alta taught at a Kindergarten through sixth grade elementary school in northeastern Philadelphia. She had taught at this school for nine years and in the district for 20 years as of 2006. In describing the school's neighborhood, Alta indicated it was a middle class neighborhood with a predominance of Caucasians (anonymous, personal communication, May 4, 2006). The school was 50 percent Caucasian, with about 25 percent African American, 20 percent Latino and 5 percent Asian students. Sixty-one percent of the students came from low income families, making them eligible for the free breakfast and lunch program. Attendance at the school was about 90 percent overall. Mobility of the students was only 13 percent, meaning the student population was rather stable throughout any given school year. Overall enrollment for this school was classified at the middle level as defined for this book.

Before Alta participated in the CPI program she did not have a computer in her first grade classroom. "There might have been a few computers in Kindergarten classrooms, and the school had a computer lab, but none of the first grade classrooms had computers. In fact, she did not recall upper grade classrooms having computers either.

My teaching, pre-CPI, was teacher-directed. I set the curriculum, pace, and all. They were just starting to implement Balanced Literacy as the literacy curriculum in first grade and I knew that computers were going to be part of that eventually. I saw one of my colleagues with a new computer and asked how come she had computers in her classroom when I didn't. [This teacher] told me about the CPI program, and I knew that this would be a way for me to achieve three goals – (1) to learn more about computers myself, (2) to learn about using computers in my teaching, and (3) to get a computer for my classroom. CPI did that and much more.

In 2006 Alta had seven computers in her classroom. They were set up as a technology center for center time. She had two programs for her students who were not reading on grade level. "There are materials there for every subject and lesson plans for every grade level."

One example of what I was doing with my students was a school wide project during the winter. The school project was a social studies project about 'the world.' Each class was assigned a country, and my classroom was assigned Australia. The students were to do research on the literature about Australia, legends about Australia and the indigenous peoples, basic information about the country, the weather, and animals. Then, they put together a performance for the school. In doing the project, we incorporated the Internet research into art work (posters to be used in the performance), writing (summarizing their findings and telling stories), math, including bar graphs about the weather, and science by comparing animals across countries. I probably would not have done such a project before because I did not have a computer in my classroom, and I would not have allowed the children to work on their own. But the Resident Teacher I visited through CPI, along with what I learned through the CPI Saturday sessions, gave me the confidence that the children in my classroom could be 'safe' on the Internet, especially if I managed their access through the kid-safe portals, like Yahooligans™, that I learned about through CPI.

With all the computing power in her classroom, and the set of skills established through the CPI program, Alta had now transformed her teaching style to more of a student-centered approach than teacher-centered. "I no longer am the source of all instruction because I allow the children to do peer teaching using what they learned independently" (Peer Systems Consulting Group, 2006). This allowed all the students to be supportive of each other, just like the Medicine Wheel they discussed in the CPI program in Chapter IV.

In spring 2006, Alta had 32 children in her first grade classroom, double that of the previous year. The room was large and the children sat at desk groupings such that four children sat across from four other children, a group of eight. There was a large rug in one area of the room with a dry erase board for whole group instruction, among other uses. There were books on racks and in baskets, a math word wall and a reading word wall. Under the windows were the computers. Due to the lack of space with so many children in the classroom, the only "activity center" was the technology center. The other centers were in baskets that were brought to the tables for the children to experience science, art and more. The daily schedule was posted for all to view, along with the daily focus skills and strategies for reading and math. Student work was posted on the bulletin board inside the classroom, as well as outside, for all to see. Celebrating student achievement was a given.

All the children got a turn at using the computers each week. There was a chart near the technology center with student names on cards. As a student got his/her turn on the computers, their card was turned over. This continued until all the names were turned over and the process began again. Students used a variety of programs to support the Balanced Literacy curriculum and the math curriculum.

The key to success was seeing a difference in student performance. Alta noted that since she had become more confident in using the computers in her classroom, and had increased the number of computers in her classroom, she had also expanded her expectations of her students. She expected them to do better in math and reading, and indeed she had noted that they were leaving her classroom at higher achievement levels in both subjects. She attributed that to the changes in her teaching style that came after participating in the CPI program.

Melanie

Melanie taught at an elementary school located in the northeastern region of Philadelphia. The school was composed of Kindergarten through fifth grade classrooms. Enrollments were classified for this book as high. The student population in 2006 was mostly white, with over 50 percent Caucasian, nearly 20 percent Asian, 13 percent African American, and the remaining students classified as either Latino or other. This population mix had changed during the last five years. In the 2001-2002 school year, the student population was 70 percent Caucasian, 14 percent African American, 10 percent Asian, and nearly 6 percent Latino. Melanie noted that the student population had become quite diverse with more Asian students and African American students. She also noted that students selected to go to this school because it had a strong academic focus and students did extremely well on various outcomes measures (anonymous, personal communication, May 11, 2006).

In describing the neighborhood, Melanie said it was a middle class neighborhood that was mostly Caucasian, but with diversity. There were more Asians in the neighborhood in 2006 than before, as well as Latino families.

Melanie had participated in many professional development programs, all to support her students. She had taught at this school for 11 years and for over 30 years in the school district starting at a school in northern Philadelphia. One professional development program she remembered was in support of Balanced Literacy just in the last few years. From that program she received reading materials and hardback books, which are always of value in early childhood programs. She also received "read-to" books, which have been extremely useful.

Melanie participated in the CPI program in either 2001 or 2002. She noted that she had taken so many professional development programs that it was often hard to remember them all. She said (anonymous, personal communication, May 11, 2006):

But I remembered CPI because it was about computers and I wanted to know more about computers and to get a newer one for my classroom. Before CPI I knew how to insert a computer disk and use the keyboard, but really knew nothing about accessing the Internet or spreadsheets. The program was an eye opening experience learning about all that is available to support my teaching and my students learning through the Internet.

We didn't have the Internet 30 years ago. We didn't have computers in schools, they were in businesses. Many of the teachers who participated in CPI were older with lots of teaching experience, and computers were not used as part of the classroom practice because it was not part of their experiences as either students in public school or in teacher education. We simply didn't know what to do with them.

Melanie noted that her teaching in 2006 was quite different from before CPI. Before CPI she rarely used the one "ancient" computer in her classroom, if at all, and never with the children. Now she used the CPI computer daily. The children used it or the older computer at least once a week. Melanie noted that the older computer was mostly used for word processing, which was fine for students to edit their writing. Melanie learned how to access lesson plans, materials, stories and activities from the Internet that she never knew existed.

I had an educational aide in my classroom for two and one-half hours a day. This time was when I often had the students use the computer. The aide helped to monitor what the children are doing. While many of the children came to school in the early grades saying they 'know how to use a computer,' it was often difficult for them when they encountered programs different from what they were used to using. So having the aide nearby to help when there were problems had been a real help to me.

Students who were less experienced used the computer for keyboarding skills and for creative projects, like illustrating a story or drawing pictures. These were young children and they loved to draw, and for some it might have been easier on a computer than with the crayons due to lack of dexterity at that age. Melanie also monitored their journal writing. As they completed a writing project and did it well, they were encouraged to use the computer to type and edit their writing, and to revise before printing out the work to be displayed around the room.

Melanie also had software to help the children with both phonemic awareness and number awareness. With Balanced Literacy, in particular, phonemic awareness was but one component of literacy, but an important one. For those children who were perhaps from families where they were either just learning English or English was not frequently spoken at home, gaining awareness of the sounds of English was essential for fluency. Computers were wonderful teachers in that regard, and very patient with young learners.

This school gained a school annex early in the 21st century, and this was where Kindergarten and first grade classes were housed. Melanie noted that her classroom was not as large as the one she previously had in the main building, but it had high ceilings and windows, thus it was a bright environment for the children. The windows faced the school yard, so there were no trees and such, but the view was still pleasant. On one wall were cubby holes where the children stored coats, book bags, and other materials. The two computers were along the back wall ready and available for use. There were six tables for cooperative learning groups, along with tables around the room for the science center, art center, and play center.

Her first grade classroom had the 100 Book Challenge®[5] set up in one area, and there was a weather chart. There were also three word walls, one each for literacy, math, and science. There were posters and children's pictures posted around the room. And they had an aquarium with a few guppies. They went to the aquarium on their class trip and made fish to hang around the room to remind them about that experience. They were also learning about butterflies, so they had some caterpillars waiting to blossom into beautiful butterflies.

Overall, Melanie noted that she would never have thought of using the old computer she had with her children before CPI. "It was old and antiquated." Providing opportunities for her students to use the computers to support their reading and science projects was an important step for Melanie who had never done anything like that before. Change is not easy for most teachers. For a teacher with over 20 years of teaching experience and with classroom teaching methods honed over those many years, adding a new component to the repertoire of teaching took courage and support. CPI was there for Melanie when she knew it was time to bring new tools into her classroom.

Summary

As with the Kindergarten teachers, these teachers described themselves as having minimal to novice skills in using computers before participating in the CPI program. Only Parker talked about his use of the older computers with his first graders. Participating in CPI was a way to get a newer computer for their classrooms, as well as to gain the skills they knew they needed to support their students as they moved into the 21st century. Three of the schools represented by first grade teachers were described as having high achieving students. While the teachers did not mention that this was an influence on their decision to learn more about using computers to support their classrooms, one can surmise that being able to use computers along with their students could have been a motivating factor. As Healy noted, "trying to use computers exposed [teachers] to potentially embarrassing situations, which could undermine either their sense of competence, their classroom authority, or both" (1995, p. 112). CPI helped these teachers avoid such compromises. Mary Jo came from a school where the CPI teachers interviewed agreed that the CPI program itself was time for them to "come into the 21st century" and use technology with their students. Several of these teachers radically changed their teaching practices and clearly attributed this to the program. Additionally, having the time to learn software, to practice new skills (e.g., creating a Web page), and collaborating with other teachers were important to these teachers.

Analysis

There are many teachers in every school; each one has his or her own classroom filled with children. There is little room during the school day for collaboration between teachers within the same grade level, much less across grades. Because the CPI program was not organized by grade level, but offered to any teacher in the district, teachers from different grades came together and learned from each other, a variation of peer coaching and/or mentoring. Even the Resident Teacher experiences for the Visiting Teachers were not always on grade level. Having the opportunity to share with other teachers about concepts and skills they were learning empowered these teachers in ways they had never experienced before. As Mary Jo said (p. 99), CPI was the first professional development program in which she had participated that made her feel like a professional.

But again, change does not happen easily. Mary Jo had the advantage of being located in one of the original CPI elementary schools where she had access to all the teachers who had originally participated in the program and she had a principal who supported technology across grades and the curriculum. This gave her many more opportunities to interact with teachers and to discuss ideas for infusing computers into her classroom, and perhaps was the reason she did not report a reduction, but an increase in use of computers when the mathematics curriculum changed eliminating the software options her students had been using. Being in the same school and both teaching first grade, Joan and Rhoda could support each other as well. And Mary Jo was in a school where several teachers participated in the program together, with the expressed purpose of using the opportunity to work together and support each other in this quest to become 21st century teachers. Perhaps the biggest change was in Alta, who went from being basically a non-user of computers in general to having seven computers in her classroom and students using them daily for meaningful activities with the curriculum.

These teachers were not all at the same level of change (Fullan & Pomfret, 1977) or adoption of an innovation (Hord, Rutherford, Huling-Autsin, & Hall, 1987), but they had met two of the factors that affect implementation of an innovation as expressed by Fullan (2001). But was this change based on the program, or on the politics of instructional technology as Huebner (1999) discusses? Were these teachers seduced by the "politically correct" perception that computers were important to classroom practice in the 21st century, or did they honestly believe that computers brought something to their practice that could not be accomplished any other way? Perhaps there was some of both in the end.

Fullan and Pomfret (1977) noted that there were four "determinants of implementation" of an innovation in a classroom—the characteristics of the innovation, the strategies for implementation, the characteristics of the adopting unit, and the demographic factors of the organization. In the reported cases the innovation was the use of computers within the classroom. All but one of these seven first grade teachers reported they classified themselves as novice computer users before they participated in CPI, and only one teacher used computers with his students "on occasion." From the CPI experience, they each changed in unique ways, especially Alta.

Implementation of computers into these classrooms was not as similar as that for the Kindergarten teachers in Chapter V. For instance, Joan very clearly noted that her classroom did not have the infrastructure or electrical supply necessary to support simultaneous use of multiple computers. This clearly put implementation into a different light than in classrooms where support for electrical appliances was sufficient. As for the strategies for implementation, the CPI program met the needs of these teachers. The training was at their level of experience (novice, as reported by most), availability of support in time and materials to become comfortable using computers for their own needs and seeing how "master teachers" guided student use of technology. Feedback opportunities were in place, and level of participation was high. Each of these teachers reported on how essential it was to have the opportunity and time to practice new skills and to consider new ideas afforded to them both through the formal training and the informal training with the Resident Teacher.

As for the characteristics of adopting, four of these teachers initially took what Fullan and Pomfret (1977) described as an opportunistic approach to participation rather than a problem-solving approach. By this I mean, they took the professional development program to get a computer or a newer computer for their classroom. While that may have been the initial reason for participation, in the end their classroom practice changed significantly, especially for Alta and Parker. As for demographic factors, each of these teachers was in a school where the students were considered disadvantaged socio-economically. They were aware that these children did not have access to computers outside of school and did not want them to be disadvantaged vis-à-vis the "digital divide" in school. They used their political capital to gain resources for their classrooms for the sake of their students (Huebner, 1999). Providing every advantage to these students was important in the long term.

In regard to the Concerns Based Adoption Model proposed by Hord and colleagues (1987), each teacher was at a different stage of adoption. Mary Jo was in a school where the teachers closely collaborated around CPI and CPI principles, which gave her the opportunity to reach the collaboration level of adoption. She was the only first grade teacher interviewed who capitalized on peer coaching and mentoring, so important for some teachers. (Ball, 1996; Berman & McLaughlin, 1975; Fullan & Pomfret, 1977; Guskey, 2000; Joyce & Showers, 2002; Little, 1989; Sharan & Hertz-Lazarowitz, 1980; Wilson & Berne, 1999) The other five teachers reached the consequence level of adoption where they were concerned about the impact of the new technology on their students, and perhaps the "digital divide." As for levels of use, it is fair to suggest that all but Alta and Parker had reached the level of routine use, but not to the point of individualized instruction. Alta, with seven computers in her classroom, could assign individualized projects, but did not mention this. Parker, as the new computer teacher in the school, was looking forward to the new Instructional Technology curriculum to become required so he would be able to use the curriculum to meet the individual needs of the children in each grade level.

Overall, these teachers embraced what CPI had to offer their classrooms in unique ways, but in the end change did occur at different levels of intensity and sustainability. Using the effective strategies of adopting an innovation posited by Berman and McLaughlin (1978), each teacher adapted their practice to accommodate what technology could bring to their classrooms. In so doing, they began a journey toward bridging the digital divide for their children's future.

References

Ball, D. L. (1996). Teacher learning and the mathematics reforms: What we think we know and what we need to learn. *Phi Delta Kappan, 77*(7), 500-508.

Barnes, B. J., & Hill, S. (1983). Should young children work with computers? Logo before Lego? *The Computing Teacher, 10*(8), 11-14.

Berman, P., & McLaughlin, M. W. (1978). *Federal programs supporting educational change, Volume VIII: Implementing and sustaining innovations*. Santa Monica, CA: Rand.

Clements, D. (1998). *Young children and technology*. Paper prepared for the Forum on Early Childhood Science, Mathematics and Technology Education. Washington, DC: U.S. Department of Education. (ERIC Document Reproduction Service No. ED416991.)

Dewey, J. (1938). *Experience and education*. New York, NY: Macmillan.

Fullan, M., & Pomfret, A. (1977). Research on curriculum and instruction implementation. *Review of Educational Research, 47*(1), 335-397.

Gardner, H. (1999). *Intelligence reframed: Multiple intelligences for the 21st century*. New York: Basic Books.

Guskey, T. R. (2000). *Evaluating professional development*. Thousand Oaks, CA: Corwin Press.

Healy, J. M. (1998). *Failure to connect: How computers affect our children's minds – for better or worse*. New York: Simon & Schuster.

Hord, S. M., Rutherford, W. L., Huling-Autsin, L., & Hall, G. E. (1987). *Taking charge of change*. Alexandria, VA: Association for Supervision and Curriculum

Huebner, D. E. (1999). Politics and the curriculum. In V. Hillis (Ed.), *The lure of the transcendent: Collected essays by Dwayne E. Huebner*. Mahwah, NJ: Lawrence Erlbaum, pp. 15-22. Originally published in A. Passow (Ed.) (1962), *Curriculum crossroads*. New York: Teachers College Press, pp. 87-95.

Joyce, B., & Showers, B. (2002). *Student achievement through staff development.* Third edition. Alexandria, VA: ASCD.

Karoff, P. (1983). Computing head start. *Teaching, Learning, Computing*, 44-50.

Little, J. W. (1989). District policy choices and teachers professional development opportunities. *Educational Evaluation and Policy Analysis, 11*(2), 165-179.

Peer Systems Consulting Group. (2006). *Peer Resources*. Retrieved on June 22, 2006 from http://www.peer.ca/profile.html

Piaget, J. (1972). *The psychology of the child.* New York: Basic Books.

Piaget, J. (1990). *The child's conception of the world.* New York: Littlefield Adams.

Pressley, M. (1998). *Reading instruction that works: The case for balanced teaching*. New York: Guilford Press.

Sharan, S., & Hertz-Lazarowitz, R. (1980). A group investigation method of cooperative learning in the classroom. In S. Sharan, P. Hare, C. Webb, R. Hertz-Lawarowitz (Eds), *Cooperation in education,* (pp. 14-46). Provo, UT: Brigham Young University Press.

Vygotsky, L. S. (1978). *Mind in society: The development of higher mental processes*. Cambridge, MA: Harvard University Press.

Wilson, S. M., & Berne, J. (1999). Teacher learning and the acquisition of professional knowledge: An examination of research on contemporary professional development. In A. Iran-Nejad, & P.D. Pearson (Eds), *Review of Research in Education*, Volume 24, pp. 173-209.

Endnotes

1 Of interest is the fact that in Pennsylvania, parents are not obligated to enroll their children into school until they are the age of 7, which means they may miss out on Kindergarten completely and be a year older than other children in first grade.

2 The DIBELS is the Dynamic Indicator of Basic Early Literacy Skills which is used with pre-school through third grade children to identify reading readiness and fluency.

3 Prior to 2002, the School District of Philadelphia was divided into 22 clusters of schools. Each cluster had an administrative office with staff for curriculum, technology, and more.

4 Balanced Literacy is defined as a combination of teaching phonics for understanding sounds of letters individually and letters in words with whole language instruction aimed at understanding text and gaining a love of reading. (Pressley, 1998)

5 The 100 Book Challenge is a unified reading program created by the American Reading Company. The intent is for children to want to read and to read on grade level. The unified system includes access to great books, a system assigned reading at the appropriate reading level, reading at school, reading at home, accountability for reading, and professional development for teachers. For more information, access http://www.100bookchallenge.com/, last accessed on June 29, 2006.

Chapter VII

Digital Drawing to Digital Graphics: Second Grade

Children need opportunities to make choices about some of their computer experiences. In addition, kindergarten or primary-grade teachers will want to use the computer for more directed activities that match their learning objectives (Haugland, 2000, para. 7).

Do you remember your second grade classroom? I remember it being a rather large room, but then I was quite small back then. There were music instruments in one corner of the room that we used to "make music" every week. There were books along one wall, and small tables in the middle of the room where students sat. Students were grouped at the tables such that there were six to eight at any one table. Projects were done together and work was shared giving encouragement or trouble. There were windows along the wall across from the entrance that looked out onto a tree lined street. The school was a Kindergarten through sixth grade in a middle class neighborhood.

Today's second grade classrooms, for the most part, look quite similar. Rooms tend to be spacious. Most students in the early grades, and through about third grade, want to please the teacher and be liked. They "buy into" the rules of schools and education, such as lining up in order, sitting still until called upon, and doing their work as best they can. They get rewarded for good behavior in many ways, like being the attendance reporter, getting to erase the board (and back in the day, this *was* a privilege), or getting to hand out or collect materials or homework. This is a win-win situation for both the teacher and the students since classroom management is something that does not have to trump learning.

As noted, the second grade classroom has a similar look and feel to the first grade classroom. There are more books and science kits. There is still the carpet in one area of the room for large or small group activities. But these children are becoming "big" students; they are experienced in school and know the rules. They are becoming more sophisticated consumers of knowledge, and therefore require more sophisticated ways to express their knowledge. Again, Gardner's theory of multiple intelligences (1999) plays out well in the second grade classroom with lots of interpersonal learning through working with others in cooperative learning groups, spatial learning through creating and thinking with images and pictures, and logical-mathematical learning through various math problems, among other multiple intelligence frames of mind.

This chapter presents six second-grade teachers who participated in the CPI program. Each teacher is quite different in terms of previous experience with computers, and two of these teachers were interviewed together and are presented together as opposed to individually. In reading their combined stories, I think you will understand why they were presented together rather than separately. What is important in these six stories is that they were all inclined to participate in CPI for different reasons, but in the end changed their classroom practices to provide new and varied experiences for their students.

Table 3 gives the reader an overview of the teachers presented in this chapter. The information provided includes how many years of teaching experience they had

Table 3. Kindergarten teacher summary

Name	Years teaching	Grades in school	Prior computer usage	CPI teacher
Sarah	15+ years	K – 8	Novice user	VT only
Sally	30+ years	K – 5	Some experience	VT only
Debra	20+ years	K – 5	Novice user	VT only
Jo	10+ years	K – 8	Some experience	Both
Gina	10+ years	K – 8	Novice user	Both
Bea	16 years	K – 4	Novice user	Both

prior to participating in the CPI program, what grades were included in their school, how they characterized their prior use of computers/technology prior to the CPI program, and whether they were a Visiting Teacher or also asked to be a Resident Teacher. This information may help the reader in deciding which teacher story is close to theirs.

Sarah

Sarah taught at the same school where Rhoda and Joan taught first grade (see Chapter VI) in northwestern Philadelphia. Again, this school was in a middle class neighborhood, but the majority of students were bussed in from other areas of the school. Enrollments for this school were classified as at the middle level as defined for this book, but there had been a loss of 25 percent of the student population in seven years. Sarah taught second grade and had taught second grade for 15 years, most of those years at this school. When asked about her experiences within the CPI program, Sarah said (anonymous, personal communication, June 3, 2005):

I didn't have much knowledge of computers before taking the CPI program. Like Rhoda, I was definitely a novice computer user. From the program I learned how to make a Web site, how to use PowerPoint and Inspiration, how to access a wealth of information about Web sites for different grade levels, along with the ins and outs of computers. I had never had a computer in my classroom before, and this was an opportunity to get at least one."

Looking at other second grade classrooms in the school, there were computers around, but the main problem for the teachers was the infrastructure of the building. As noted earlier, these older school buildings were not built with computers, or technology in general, in mind. Indeed they were built long before computers were invented in the large scale, mainframe versions, much less the desktop variety. And, schools where students were not from the neighborhood had difficulty getting parents involved in the Home and School Association, which might work with the schools to obtain resources for classrooms. Clearly this was a disadvantage for this school. However, as Sarah noted, the school district was doing its best to provide opportunities for teachers to learn about technology in general and how to integrate technology into their daily activities painlessly. Sarah noted that changes occur quickly with computers and the curriculum in the district. She specifically said:

I am still learning a lot, but I'm much more comfortable with computers than before. I feel like I can allow my second graders to use the computer and the Internet safely.

I do reports, my lesson plans, grades, and enter DIBELS [see Chapter VI for a description] data online. My second graders are using First in Math® to gain math skills, and doing more writing on the computer than before. But one plus about the standardized reading curriculum in the district is that the Web site for the text has materials to supplement the reading. The children love these activities, and I would never have been able to access them without taking CPI. I know I wouldn't!

While there are some critics who would point out that Sarah has not actually "changed" her teaching practices, she now had confidence in using a resource to which she and her students had access, and she was willing to allow her second graders to use the Internet. Through CPI training, she learned safe ways to allow young children to do research and experience the world through kid-friendly portals on the Internet, something she would never have imagined doing before CPI. In addition, the publisher of the reading curriculum provided supplemental materials, exercises, and activities. Before CPI, Sarah noted she would never have taken advantage of these resources because she was afraid to use the computer at all. In that respect, Sarah had changed her teaching practices to allow her students to explore the resources found through the kid-friendly Internet portals she made available to them as well as the reading curriculum materials.

Change does not need to be wholesale, but can be incremental, leading to other opportunities. Having 10 other teachers from her school participate in CPI was also helpful for Sarah. While they were from different grade levels, they were all in early grades and could share and be peer coaches or mentors for each other. Having a cohort of teachers from the same school was a benefit to each and every one of them (Ball, 1996; Berman & McLaughlin, 1975; Fullan & Pomfret, 1977; Guskey, 2000; Joyce & Showers, 2002; Little, 1989; Sharan & Hertz-Lazarowitz, 1980; Wilson & Berne, 1999)

Sally

Sally taught second grade at a school in the northeastern region of Philadelphia. She had been teaching at her school for six years, but had taught for 30 years total. Her school was the same school where Ronnie taught Kindergarten (see Chapter V), so she had high achieving students in her classes even in second grade. The school enrollments were classified as low level for this book, and had increased by six percent since 1999.

When asked about use of computers before she participated in CPI, Sally said she had a computer at home and had used computers since the 1980s (anonymous, personal communication, June 6, 2005). She had a digital camera and used LOGO and Pascal programming languages previously. She felt somewhat confident in her use of computers. She already knew how to use the digital camera for images and how to burn CDs. She had a media-PC at home and was still working on how to save to DVD. She used the Internet for games. She described herself as a confident, high end user of computers for personal use. Clearly compared to the other teachers discussed so far, Sally was a "high end" user of technology, so why did she participate in CPI?

Sally participated in CPI to get a computer for her classroom and to learn more about the Internet. CPI made her feel more comfortable with the Internet, made it easier to use because she was not afraid of it and now had a list of Web resources perfect for her second graders. She still felt she had difficulties with some things, but did feel mostly comfortable.

I use Headsprout.com™ with my students for phonics and sounds. My classroom has a computer center with four computers. It is set up so children using the computers can focus on their work and other children can focus on what they need to do. This is something I learned through CPI, how to have multiple activities happening in the classroom successfully. I would never have thought of or tried that before CPI. We were lucky to have a Resident Teacher here at our school, so we could go over and observe her any time we had a prep period. That was a true bonus for us.

The main benefit for Sally from CPI participation appears to be learning how to have multiple activities happening in her classroom at the same time. While she would never have attempted this before CPI, she found that her second grade students were capable of doing different tasks and activities and being productive. She learned that having students doing different activities simultaneously did not cause the chaos and confusion she thought it would.

Sally also had the advantage of having a Resident Teacher in her school, which meant that she did not have to totally depend on the visitation days to observe and gain from the experiences of other teachers. As with Sarah's school, this school had 11 teachers participate in the CPI program, which established a cadre of teachers who could call upon each other for support. They were in different grades, but that did not stop them from being supportive across the school.

Debra

Debra taught at the same Kindergarten through fourth grade school as Marta (Kindergarten, see Chapter V). She had taught at this school for over 20 years. Several teachers started with CPI in 1999 and participated together. As noted before, these teachers all had the same lunch period so they came together at that time to talk about CPI, to share experiences and ideas, discuss resources, and more. The school was in the northeast region of Philadelphia.

This school, in 1999, was described as an urban, small, very close knit school with no turnover in teachers. The students were very diverse ethnically (African American, Hispanic, Asian, among others) and high performing on high-stakes tests.[1] Debra said there was no difference between children's performance on standardized tests at this school based on race or gender or ethnicity. They all succeeded and performed highly on the tests, like the nationally norm referenced tests used in the district in 1999. There was a lot of pride in the school, with a very professional and enriched staff. Communication lines were open between all teachers and the staff, according to Debra. Teachers read a lot of literature on teaching methods and talked together about teaching. Debra said (anonymous, personal communication, December 10, 2004):

I wanted to learn how to use computers and CPI was an opportunity to get supplies and materials for my classroom. Most of us [at this school] considered ourselves as 'computer illiterate.' The Home and School Association didn't have much money, so we didn't have a lot of computers in the school. With CPI, we could get the technology for the classroom. And, we wanted to learn how to use a computer, period.

In those days, if you got a computer, it arrived in the box at your [classroom] door. But we didn't know how to set it up or to use it. In 1999 there was one computer person in the school we went to for help.

I had used the Apple IIE previously with my Kindergarten class. I learned from my own children at home, not my students, how to use AppleWorks and PrintShop®.

Unlike many teachers who participated in CPI, Debra was not afraid of technology, nor did she really consider herself a neophyte. She had learned through her biological children how to do some activities through a computer. She may have described herself as "illiterate," but she already had basic skills for personal use. Debra just had no idea how to incorporate computers into her classroom. She had one computer in her classroom before CPI, but it was still in the box because she did not know how to set it up and was afraid she might damage it doing something wrong. Tech support was not readily available in the school with only one person for the entire school.

The CPI tech support was unbelievable. The CPI Director, in particular, was wonderful. When you called, she came right out. Also for independent work, children could start a project in the classroom, work on it in the computer lab, and then come back to the classroom and present to the class. It was great for the children. Once you got teachers comfortable with what they were doing [on the computer], that made computers accessible for the children. Before CPI, we teachers didn't use the computers, so the children didn't. Now we are dependent on the computers! We have between four or five computers in every classroom now.

The principal was extremely supportive. She allowed us time to practice, which is critical to retaining our skills. When I forgot how to do something, I would call "help" and the children would help me. More of the children, as they got computers at home, were great resources in the classroom.

One of the most important aspects of CPI was the support provided by the program director and the Resident Teachers. Teachers needed support when learning to use computers because they just did not have the time or knowledge when things did not "go right." Accessibility and support were keys to teachers using computers. Debra added that the CPI Director would come to the school to help any time they had a concern, which really gave them a sense of personal support they had never gotten through any other professional development opportunity.

Debra shared that the principal had been at this school for several years and was very supportive of the teachers and their participating in the CPI program. She shared that the principal encouraged them to participate in the CPI program as a group so they could support each other, and become resources for the other teachers. The fact that this is a small school and so many teachers participated (a total of 10), the fact they had a common lunch where they could share their ideas and experiences, and the fact that the principal was so supportive all made a difference for these teachers.

Having a supportive principal resonates again with what was discussed by Arlene Kramer and Jerry Snyder (see Chapter I). Without a principal who understands what is required for a teacher to change practice, classroom practice is not likely to change. In addition, having a group of teachers who worked together as peer coaches was very beneficial in this school where technology support, overall, was limited.

Jo and Gina

Jo and Gina participated in the CPI program in 2000. They were interviewed together, and, as you will see, their stories are so intertwined as to be difficult to present separately and do justice to both. Jo had been teaching at this school for less than 10 years, while Gina had been teaching there for over 10 years. Both Jo and Gina eventually became Resident Teachers.

The context for both of these teachers was the same school in upper north Philadelphia where Ellen teaches Kindergarten (see Chapter V). When they participated in the CPI program, they both taught second grade. All three teachers were in the same school annex at this school, which housed Kindergarten through second grade classrooms. The newer building afforded them resources probably not available in most of the schools or school buildings built before the late 1990s, except in other school annexes built at schools where the student populations were growing.[2] English as a Second Language (ESL) was a major component of offerings at this school across all grades. Because of the location of the school, many different languages were spoken by the students and their parents.

All the classrooms in the "little school house" were very large, with ample room to support lots of activities as well as computers. Internet hookups were on both sides of the rooms with enough electrical outlets to support anything needed. This building was only four years old in 2004, and they moved into it just before participating in the CPI program. The space and infrastructure allowed these teachers to experiment with configurations in the classroom so there were computer centers in more than one area of the classroom. Clearly, this would not be possible in most schools or classrooms.

Jo was in a regular second grade classroom with 30 students when she participated in CPI in 2000. She described the school as a multicultural school with 48 different nationalities in the school (anonymous, personal communication, November 3, 2004). Economically, 80 percent of the children were disadvantaged and were eligible for free breakfast and lunch through Title I. In a class of 30 students, Jo estimated that maybe 10 percent of those students had a computer at home.

Jo and Gina were best friends. In fact, Jo talked Gina into going with her to the CPI program where Gina became a technology convert. Jo put it this way (anonymous, personal communication, November 3, 2004):

The CPI program was looking for a second group of CPI teachers and was recruiting teachers outside the original four schools. The computer person from the cluster[3] was inviting people to participate from around the cluster and no one was taking it up. The cluster computer teacher asked me if I might like to participate because she knew I was relatively new to the school, willing to take risks, and already interested in computers. So I decided this sounded like a good idea, especially since I would get a computer for my classroom. I encouraged Gina to go along too. Gina, on the other hand, thought this was a crazy idea because she didn't use computers for much of anything at that time. We participated in a two-week summer program where we observed teachers at [the elementary school] working with summer school children."

Gina was in a regular second grade classroom with 30 students, as well. At that time, she did not use computers for anything. She felt she could do just as well with paper and pencil and did not need the aggravation of the computer. Gina said (anonymous, personal communication, November 3, 2004):

We went as Visiting Teachers first to the summer program. At that time, we had nothing (e.g., no computers) in our classrooms. Getting a computer was a big draw to participate. Also getting software was a draw. We got software bundles for our classrooms. Being able to learn things we didn't know before was great. We learned from the Resident Teacher about getting on the Internet and sites where our children could do research in a safe way, like the Little Explorers™ Web site. We actually saw it working and we said, 'we can do that.' Then we went back to our school and adapted what we had learned to our perspective. The computers were a big pull, but we got more each year we participated as Resident Teachers – the large screen TV, a digital camera, a video camera. These were great incentives. We are now full of technology and all through CPI (Gina).

Jo further noted that the principal played an important role in allowing many teachers from their school to participate in CPI. In the end approximately 10 teachers from the school became CPI teachers and, subsequently, became resources for others in the school as described by Ellen in Chapter V. Jo said:

The principal was very supportive. She was the same principal now (2004) as back then. When Visiting Teachers from our school were ready to do a visiting rotation, but there was no substitute for that classroom, the principal did not hold them back. We covered that classroom. The principal also learned a lot about technology from all the teachers participating in CPI. One year we came back with a digital camera. Right away she thought this would be great for the school, but one was not enough since there are two buildings. So, she went out and bought digital cameras for the school. She understands that software is important, and she has supported us because she sees changes in teaching and student learning.

They participated initially in the CPI summer program at the first CPI elementary school as presented in Chapter IV. Gina described it this way:

The summer program included training at the central technology offices, then to the elementary and middle school where CPI was first developed. At the end there was a big fair at the middle school to see what the children put together. At central offices, there was training, even in simple things like how to turn on the computer. Some people didn't know those things. The CPI Director was the motivation behind the whole thing. She taught the course and made you feel like you could take on this project. She put you in a comfort zone where you didn't feel stupid.

Jo had been using computers for a while, but Gina had not. In fact, Gina never used them. "When you don't use something, you have a fear of it. I really didn't need to use a computer. I was fine with pen and paper. Or when there is a glitch, like you go to do something and the computer doesn't work the way you want it, the reaction is 'this is stupid.' I can use the paper and pencil and I won't be stupid." She went on to describe her transformation from non-user to constant user like this:

I use them all the time now. I never, ever, ever used them before; except when I taught eighth grade, I would have the students do word processing. But I only had one computer, so it was easier to have them do word processing in the computer lab [with the computer teacher]. But after CPI and learning about all the things available to us and getting the big screen TV, it became simple. The training showed us [Internet] sites they used. And it was watching [Resident Teachers] use the computer in the classroom and not have chaos and craziness in the classroom that convinced me to change my teaching strategies. They were doing scavenger hunts and using the software 'Stories and More.' We went for a couple of days visiting. At the middle school we saw older kids making up flyers and movies. We put it all together and knew our kids could do this.

They both noted the different types of software options the CPI program opened up for them. While both of these teachers were veteran teachers, they did not have a computer class in the teacher training programs. Imagining what computers and grade appropriate software could bring to their practice was difficult for them. One of the programs to which they were exposed was ClairsWorks for Kids® which opened up avenues for many options. Jo noted:

One of the things we learned was ClarisWorks for Kids. We used this for journal writing. When the children saw the paper lines like the paper we used, and they saw the letters in print, they loved it. Then they could draw a picture. They had a sentence they had typed and a picture to illustrate the sentence. It was eye-opening. That's where we started—word processing and computer graphics. Then we did a whole lot of Internet searches using Little Explorers from the Enchanted Learning® Web site[4] *and more.*

With only one computer in classrooms, the tendency was for only the teacher to use the computer. Gina discovered that, even without a monitor to enlarge the screen, she could engage her students around the computer to achieve higher learning goals. Gina described this activity:

I used to have the 30 children sitting on the carpet in the classroom with the one computer, no big screen TV. The children would have a copy of the quiz from Little Explorers and we would do the research together. They would find the letter, and scroll down to find the information. It was great.

Students learning and making learning much more fun for them was what it was all about. They, the students, had to figure out how to use the scroll bar to find information. There were also all the words in between the start of the list of words and the word they were searching for, and they would find something that intrigued them and click on that word. They learned about things they didn't know about before. It gave them a lot of background knowledge than they had at that time. This really helped with the ESL population because of the pictures. They might see the image of a 'mouse' and then see the English spelling of the word and could make the connection and learn the words better and sooner.

As noted before, participating in the CPI program was truly a transformational and enlightening process for Gina. "I went from not using the computer at all to using it every day. They [her students] use Kidspiration® to create their Web sites. Their journaling is on there. I was the Kidspiration demonstration room when teachers came to visit."

Eventually there were four teachers from this school who were Resident Teachers. In 2004, Jo and Gina were the only two of these four teachers still teaching at the school. The four teachers would get together and plan for visits by CPI Visiting Teachers. Jo noted:

We planned for a Visiting Teacher to visit all the classrooms. You don't use the computer all day long, so when one classroom was doing something that didn't involve a computer, the Visiting Teacher would rotate to another classroom. They saw Kidspiration in Gina's classroom, and they saw something else in my classroom. We made sure the teachers saw a variety of activities. For instance, Gina only had one computer, while another teacher had two computers and I had several computers as a small learning community coordinator. They got to see the one computer classroom where the use was more teacher directed, the two computer classroom which was used for center time, and independent use in the classroom with several computers. In the end, what people learned was not that the computer is the most important thing, but how to manage the classroom and get the most out of the computer. By seeing this in action, they could see how it would work in their classroom.

This group of Resident Teachers planned for a Visiting Teacher's visit. They fully realized that during the school day, computers would not be used all the time in every classroom. There were other activities students engaged in that were equally important and meaningful, but the Visiting Teachers were not there to experience recess or lunch, to name a few. So creating a plan for these Visiting Teachers to specifically experience a one computer classroom, a two computer classroom and a multi computer classroom gave the Visiting Teachers a richer experience in the long run. They would have multiple examples of how to change their classrooms.

Gina thought participating in CPI to learn about computers was a total waste of her time until she got into the training and saw the possibilities with children in early elementary grades. Jo was willing to try new options, but Gina needed to be convinced. In the end, both of these teachers changed their classroom practices, allowing their students to benefit from what computers could bring to learning experiences that could not previously be accomplished. Being able to engage all 30 children at the same time with a small computer screen is not easy, but Gina was able to do it. Even as they became Resident Teachers within the program, they counted on the students to be ambassadors for how computers could support learning and change classroom practices.

Bea

Bea taught at a Kindergarten through fourth grade elementary school located in northern Philadelphia. When she was interviewed in 2006, she had taught for 16 years in the district. She taught at this school for 10 years and really enjoyed being part of the school's family. The school was rather large for a Kindergarten through fourth grade school with a high enrollment, as defined for this book. The enrollment had increased by over 25 percent since 1999. Nearly 70 percent of the students were classified as from low-income families, which was just 2 percent below the district average. Approximately 60 percent of the students were African American, with 17 percent Asian, 15 percent Latino or Hispanic, and the rest of the student divided between categories of Caucasian and other. Student attendance was at 93 percent which was 3 percent better than the target for elementary schools in the district. Turnover in the student population was only at 13 percent, significantly down from the 26 percent rate of 1999.

When she participated in CPI in 2001-2002, Bea had what she called a "very old computer" in her second grade classroom. "It didn't work very well. The children played one or two different games on the computer, but it really wasn't good for much of anything else. There wasn't much software available for such an old computer," Bea said (anonymous, personal communication, June 13, 2006).

Bea found out about CPI when the principal sent around a flyer to let teachers know about the program. There were 11 teachers from this school who participated. "We were interested in getting computers for our classrooms. And I was interested in learning more about computers and this seemed like the right program." In the end, Bea received a new computer and a printer from CPI, along with various software programs for second graders and phonics games. She learned how to make slide shows and a home page on the Internet. She said she used slide shows with her students when appropriate.

The [CPI] course was good. The director was terrific. It was the first time I had used a laptop, and having the laptop was wonderful for doing the projects and homework. After the program I bought an iMac computer for myself and brought my old computer to school to use there. I gave my home computer to the classroom so I had two computers in the classroom along with the very old one that was there to start.

And visiting other teachers [the Resident Teachers] was great. The teacher I visited showed me how to use slide shows with her students and how she set up a rotation for children to use the limited number of computers she had in her classroom. The children knew how to rotate on and off the computers quite well. They knew when they were to use the computer, what they were supposed to do there, and how to save their work. When a child finished her turn on the computer, the next student came over. It worked like clockwork. I knew I could do this in my classroom, depending on the students.

One year Bea had a student in her class who really took to using the computer. This young lady worked on stories and inspired other students in the class to write as well. She was skilled with the computer and stimulated a desire by other students to do their work on the computer. From this beginning, Bea started an "Authors' Party." The children wrote stories, typed them up and illustrated them on the computer. Bea printed them out and made books of the stories for the children. That year they published two books. Although she had a printer in her classroom before she took CPI, she and the children did not use the old computer, so they also did not use the printer. Creating the story books was something they could not have done without the resources made available to them through the CPI program plus the ideas she got from other CPI teachers.

All in all, Bea changed her classroom teaching to include computers as part of her teaching. She even bought an iMac for home and took her older home computer to school to use with her students. Clearly, one computer was just not enough for Bea. While she did not create a lot of slide shows to use with her lessons, she encouraged her students to create slide shows to demonstrate their learning and new knowledge. Creation of the Author's Party to help the students develop stories for books was something she could not have done before CPI basically because she did not have the tools, but also because she could not imagine such a project.

Summary

Looking at these six teachers, much more variety was noted between their past experiences and that of the Kindergarten or First Grade teachers from Chapters V and VI. Sally had programmed in LOGO and Pascal and was very familiar with computers;

Debra knew that she needed to become more familiar in order to help her students be 21st century citizens; Jo was comfortable with computers, but wanted to know more about using them with children; and Gina did not see a need for computers in classrooms at all but became a staunch convert after her experiences with CPI, including becoming a CPI Resident Teacher. Bea had some knowledge of computers, but considered herself a non-user. All credited the CPI program with giving them the necessary skills and confidence to bring computing into their classrooms in ways they would never have dreamt of before.

Analysis

For these second grade teachers, participation in the CPI program afforded them the opportunity to explore new ways to engage their students in authentic learning activities. Dewey would have considered these activities connected to real life since computers are completely part of life in the 21st century (Dewey, 1938). In fact, these activities were more likely to address the needs of all students in the classroom considering Gardner's (1999) theory of multiple intelligences. These teachers seemed to embrace change in their practice rather than shun it, even if Gina were a bit skeptical in the beginning. They were also willing to learn from the students, as well as other teachers, and were willing to give their own students responsibilities to help other students in their classrooms.

Looking at these six teachers through the lens of Rogers' diffusion of innovations theory (1962, 1972, 1983, 1995, 2003), these teachers did not all start out in the same stage. Sally had crossed through the knowledge stage regarding personal use of computers, but was just approaching the persuasion and decision stages for use of computers in her classroom. Debra too had experience personally with computers, thanks to her family, but none with her students. Sarah, Debra and Bea considered themselves novice users, but all three understood that they needed to be more computer literate to help their students connect computing with their own work. Gina was the one second grade teacher who needed the most persuasion. She was just approaching the knowledge stage and was pushed by Jo into seeking more information. In the end, all six reached the implementation stage of adoption of computers into their classrooms.

From the discussion by Fullan and Pomfret (1977) and later by Fullan (2001) about change in schools, changes occurred in these teachers' classrooms beyond the change in their roles and behaviors, knowledge, and understanding. Gina went into the CPI program skeptical about the value of computers in her second grade classroom. While observing other elementary grade teachers during the summer program, Gina had an epiphany about the relevance of computers in early elementary grade classrooms and that her students could do the kinds of projects and learn at the same pace as the students in the summer CPI program. Gina was the teacher who changed the most of all these second grade teachers, and in turn changed her classroom the most.

Sarah was primed to use computers with her students given her personal experience with computers. Debra was ready, as well, again because she had experience with her own son and daughter. Jo was already willing to experience new methods to support her students' learning before CPI came into her life. Gina was perfectly happy with her chalkboard and chalk, paper and pencil, and chart paper and markers. Her transformation from a non-user to a daily user herself and her new-found expectation that some experiences were enhanced through interfacing with computers were evidence of the power of the CPI model in changing teacher practice.

Unlike the Kindergarten teachers, only Debra spoke about the outside support provided by the CPI program director (see Chapter IV) and that of the Resident Teachers, per se. Sarah, Debra, Jo, and Gina all talked about the power in the number of teachers from their schools who participated in CPI, thus providing additional support the program did not provide explicitly in peer coaching and mentoring. As noted in Chapter III, peer coaching or mentoring could further support change in teacher practice long after the professional development program was over (Ball, 1996; Berman & McLaughlin, 1975; Fullan & Pomfret, 1977; Guskey, 2000; Joyce & Showers, 2002; Sharan & Hertz-Lazarowitz, 1980; Wilson & Berne, 1999)

Jo, Gina, and Bea talked about having a number of Resident Teachers in their schools and getting together to plan how to maximize Visiting Teachers' time and experiences in their schools. In almost all cases the principal was a major influence. Again, having a supportive principal made the difference by allowing time to practice new skills, providing time for teachers to get together to discuss practices and ideas, and overall establishing and environment that supported the teachers in all they did for their classrooms and students. While Bea did not mention her principal as being a driving force, her principal was supportive of 11 teachers' participation through CPI, meaning he would be supportive of those teachers being out of their classrooms for two to three days while they gained experiences from Resident Teachers elsewhere. As in Chapters V and VI, the principal was a key player, once again supporting the theory posited by Arlene Kramer and Jerry Snyder in Chapter I.

Lastly, using the Concerns Based Adoption Model (Hord, Rutherford, Huling-Autsin, & Hall, 1987) as an analysis lens, we can see that in each instance these teachers initially were at different stages of adoption and use. Sarah, Debra, Gina, and Bea were at the orientation level and the informational stage of adoption. Sally and Jo were at the mechanical use level and the personal stage of adoption. In the end, all six of these teachers progressed through the management and consequence stages of adoption and to operate at the collaborative stage with others in their schools, and they each had made the use of computers within their classrooms routine. That said, none of these teachers had moved toward individualizing use of their new computers with their students or establishing individualized instruction, nor had they fully integrated use throughout all subjects taught. Cuban (2001) would argue that since these computers were not being used every day for every subject that the change in practice was not significant. However, when you consider someone like Bea who had a very old computer in her classroom that simply did not support the curriculum, a computer that was too old to access the Internet or World Wide Web resources, getting computer resources that allowed her to incorporate them into lessons when she could put her light years ahead of where she was before. She could not have included computers in meaningful ways before CPI, even if she had wanted to and knew how. Having the resources and infrastructure to allow computers to have an impact is so important and is often overlooked in the zeal to point out that expensive technology is not being used every minute of every classroom day. It is not merely whether the computers and printers are used that is important, but whether they are used in pedagogically meaningful ways that support student learning.

It is useful to note that the teachers with the least teaching experience were Jo and Gina, and that all these teachers combined had 106 years of teaching experience between them. Changing teaching practice is not something that happens easily, as discussed previously in Chapter II. While computers have been in schools, or at least most schools, for many years, and have been part of the teacher education process since the late 1980s, only Jo and Gina, and perhaps Sarah, were likely to have experienced some formal instruction in educational computing in their teacher education programs. Unfortunately, much of such instruction before the mid 1990s was about how to use a computer (e.g., how to turn it on, how to format a disk, how to load software, how to save files in different formats, and how the applications worked) rather than how to manage a classroom where computers were part of the ongoing practice of the classroom. Further, it is important to remember that the World Wide Web has only been in existence since 1993. The world was not accessible to classrooms in the late 1980s and early 1990s in the way it is today. Many people who were teaching computing to education majors came from an educational media background where the emphasis was on creating materials to use in the classroom, not on having the students take advantage of the medium for constructing or demonstrating knowledge. Practice in higher education changes slowly as well.

Conclusion of Section II

Healy (1998) reminds us that between ages two through seven, seven types of learning need to be mastered (pp. 207-215).

1. Learning in social context, where exploring socially is better than alone.
2. Learning to use all the senses, with the essential need for intersensory integration.

3. Learning to be a powerful learner, manipulating objects actively, not as a passive consumer.
4. Learning to pay attention and to direct one's own attention, not through artificial stimuli.
5. Learning visual imagery and memory.
6. Learning to think logically through if-then situations.
7. Learning new symbol systems, like letters and numbers.

She also notes that these young children, and perhaps children who have not yet matured to age seven mentally, need adults to help make sense out of what they "see," or to differentiate between real and fantasy. She hypothesizes that, since young children cannot tell the difference between reality and fantasy, extensive interaction with technology may preclude their ability to differentiate. These 19 teachers used computers judiciously to support learning and not to replace teaching. Many critics of the use of technology with young children make the assumption that the child will be "put on the computer" and left on their own to "use it" without considering that teachers do know better.

These 19 teachers taught in vastly different schools and areas of the school district. Most of them came into the CPI program considering themselves as novice computer users, meaning they could do what they needed to do for their own work, but they were not able to use computers to support their classroom teaching or their students' learning. And there was Gina who thought computers were totally unnecessary and she could do just fine with paper and pencil. Then came the CPI program, the experiences with other teachers, the new ideas about managing their classrooms, and, for a few, the opportunity to become a Resident Teacher, of which six of these teachers did. After the CPI program, they expressed a level of knowledge and comfort with computers and using them with their students that they credited only to the program. They came away confident in their ability to cycle students on and off the computer to accomplish curricular goals, to support student learning objectives, and to maintain classroom management so all children could learn without being compromised. While the idea of "cycling students on and off the computer" may seem a bit pedantic to some, this idea was revolutionary for these teachers and gave them a sense of freedom in their classrooms allowing them to expand opportunities for learning and expressing knowledge for their students. This is not something they took lightly.

Gina's story is most compelling because she was, basically, talked into participating by her friend, Jo, who thought it was a good idea. She did not see the need for computers in early childhood classrooms and was quite comfortable without them. As she put it from her biases before CPI, "… or when there is a glitch, like you go to do something and the computer doesn't work the way you want it, the reaction is 'this is stupid.' No teacher wants to look stupid in front of their students." Gaining confidence and knowledge empowered Gina to not only become a daily user of a computer with her second graders, but also to become a Resident Teacher showing CPI Visiting Teachers how to manage a one-computer classroom. She was no longer afraid of looking "stupid" in front of her students and, in fact, established student experts in her classroom to support her and the other students. Schofield (1995) noted that there have traditionally been two roles in classrooms: the teacher and student. "An important part of the teacher's role is the exercise of authority, that is, legitimated power over the student" (Bierstedt, 1970, as cited in Schofield, 1995, p. 114). These teachers became comfortable in sharing roles with students and allowing their students to support one another.

What was important for each of these teachers was the opportunity afforded to them through CPI to spend time in a Resident Teacher's classroom. Getting away from their own classroom's daily routine to see how a colleague in a similar classroom was successful in integrating computer options into daily routines was very powerful. While the training these teachers received in the Saturday sessions or during the CPI summer program in the late 1990s was enlightening, seeing really is believing. Getting a computer for their classroom was often the initial goal for participating in CPI. However, once they saw the power in a real classroom with children who were excited about what they were learning and who were showing these teachers what they could do with technology, these teachers came back to their schools empowered to change their classrooms and to expand the traditional classroom roles so students could be teachers and teachers learners.

References

Ball, D. L. (1996). Teacher learning and the mathematics reforms: What we think we know and what we need to learn. *Phi Delta Kappan, 77*(7), 500-508.

Berman, P., & McLaughlin, M. W. (1978). *Federal programs supporting educational change, Volume VIII: Implementing and sustaining innovations*. Santa Monica, CA: Rand.

Bierstedt, R. (1970). *The social order*. New York: McGraw-Hill.

Cuban, L. (2001). *Oversold and underused: Computers in the classroom*. Cambridge, MA: Harvard University Press.

Dewey, J. (1938). *Experience and education*. New York: Macmillan.

Fullan, M. (2001). *The new meaning of educational change.* Third Edition. New York: Teachers College Press.

Fullan, M., & Pomfret, A. (1977). Research on curriculum and instruction implementation. *Review of Educational Research, 47*(1), 335-397.

Gardner, H. (1999). *Intelligence reframed: Multiple intelligences for the 21st century*. New York: Basic Books.

Guskey, T. R. (2000). *Evaluating professional development*. Thousand Oaks, CA: Corwin Press.

Haugland, S.W. (2000). *Computers and young children*. Washington, D.C.: Office of Educational Research and Improvement. (ERIC Document Reproduction Service No. ED 438926)

Healy, J. M. (1998). *Failure to connect: How computers affect our children's minds – for better or worse*. New York: Simon and Schuster.

Hord, S. M., Rutherford, W. L., Huling-Autsin, L., & Hall, G. E. (1987). *Taking charge of change*. Alexandria, VA: Association for Supervision and Curriculum Development.

Joyce, B., & Showers, B. (2002). *Student achievement through staff development.* Third edition. Alexandria, VA: ASCD.

Little, J. W. (1989). District policy choices and teachers professional development opportunities. *Educational Evaluation and Policy Analysis, 11*(2), 165-179.

Rogers, E. M. (1962, 1972, 1983, 1995, 2003). *Diffusion of innovations*. Glencoe, IL: The Free Press.

Schofield, J. W. (1995). *Computers and classroom culture*. New York: Cambridge University Press.

Sharan, S., & Hertz-Lazarowitz, R. (1980). A group investigation method of cooperative learning in the classroom. In S. Sharon, P. Hare, C. Webb, & R. Hertz-Lawarowitz (Eds), *Cooperation in education,* (pp. 14-46). Provo, UT: Brigham Young University Press.

Wilson, S. M., & Berne, J. (1999). Teacher learning and the acquisition of professional knowledge: An examination of research on contemporary professional development. In A. Iran-Nejad, & P.D. Pearson (Eds), *Review of Research in Education*, Volume 24, pp. 173-209.

Endnotes

1 High-stakes testing in Pennsylvania includes the Pennsylvania System Student Assessment (PSSA) that is a norm-reference test relating to state standards of performance in reading, writing, mathematics and science, and is one measure used to determine whether schools meet the NCLB adequate yearly progress (AYP) targets. In addition, in the Philadelphia School District, the TerraNova tests are used as nationally norm-referenced tests.

2 Again, before computer technologies came into classrooms rather than only in computer labs, the technical infrastructure such as electrical outlets, as noted before, was at a premium and not strategically located in classrooms. The newer "little schoolhouse" buildings were all built at the beginning of the twenty-first century and were all built with technology in mind.

3 This district, being as large as it is, grouped schools into clusters with a high school at the center of the cluster and all feeder schools, elementary and middle schools, included in that cluster. The intent was to have smaller "districts" within the larger district for more local support and administration.

4 This site could be accessed at http://www.enchantedlearning.com/Dictionary.html in 2005.

Section III

Teachers Voices: Elementary Education Grade Teachers (3-5 Grades)

Moving away from the early childhood grades, we find teachers who say they can discuss concepts more easily and in more depth with their students because the students are more mature and have a wider view of the world. In addition, the classroom activities move toward a combination of hands-on activities with discussion of ideas and exploration of concepts. The computer activities move away from mostly painting, drawing, and learning letters to expressing concepts through various means, including using the Internet for research for information and presentation or multimedia programs to demonstrate knowledge in unique ways. In some cases, these students are more Internet-savvy than their teachers. As with the teachers in the early childhood classrooms, these teachers were exposed to new ways of thinking about how to manage student learning through the use of computer and Internet solutions. These three chapters present six third, six fourth and seven fifth grade teachers who participated in the CPI program.

As with the teachers from Kindergarten, first and second grade classes, these teachers were asked to respond to four basic questions listed in Section II Introduction (p. 78). The stories of these 19 teachers will begin with some background information about the school and context in which they teach to set the scene. Throughout these chapters, the names of the teachers are pseudonyms to protect the identities of these teachers who volunteered to be interviewed for this book.

Chapter VIII

Inspiring the Learning: Third Grade

Although much of the teachers' trepidation about using computers in their classrooms appeared to stem from concern about how their relative lack of expertise would influence their position in the classroom and their ability to effectively and efficiently teach students using them, another factor also seemed to contribute to their reluctance. Some teachers felt quite anxious about the mere idea of having to learn to use an unfamiliar and rather intimidating machine (Schofield, 1995, pp. 115-116).

The third grade classroom is beginning to look like a typical elementary school classroom. Desks can be arranged in many ways—in groupings to allow collaborative learning or individually to promote individual achievement, or a combination of both to serve the needs of learners. Third graders are much more mature than early grade students. They know their letters. They read more complex materials with plots that are more in depth and interesting. They can read poetry and begin to understand symbolism, irony, metaphors, and similes. They can discuss ideas and present their own ideas as arguments.

Gardner's theory of multiple intelligences (1999) is extremely important in third grade. Students are beginning to demonstrate their preferences for learning different subject matter in different ways—through social activities (interpersonal), through discussions with others (linguistic), through the use of numbers and reasoning (logical-mathematical), through images (spatial), and through self reflection (intrapersonal). Each of these ways of knowing and learning, in addition to music, nature and bodily movement, are important ways for young children to experience life and learning. Not that they have been absent in early grade classrooms, but with higher order thinking skills becoming even more evident with third grade, tapping into these different intelligences allows teachers to focus on each student's needs in different ways.

At the same time, allowing these youngsters to construct knowledge based on activities situated within their zone of proximal development (Vygotsky, 1978) is an important concept for allowing them to see the relevance of subjects and topics to their own lives. Making the learning of mathematics, science, and social studies, along with reading, relate to what they know best brings the topics and knowledge home allowing them to take ownership. Computers allow students to do just that through presentation of knowledge in multiple ways that complement the curriculum and support other ways of demonstrating learning through projects, plays, dioramas, other informal assessments along with formal assessments, or tests.

This chapter presents six third grade teachers from various parts of the City of Philadelphia. They represent different ages and experience in teaching. All-in-all, they were moved by the CPI experience to try ideas out in their classrooms that, like the teachers presented in Section II, they never would have imagined. As Weizenbaum said, "Man can create little without first imagining that he can create it" (1976, p. 18). This is the same with teachers using computers in teaching. They will not use computers unless they can imagine how these machines can benefit

Table 4. Third grade teacher summary

Name	Yrs teaching	Grades in school	Prior computer usage	Overall change	CPI teacher
Mickie	10 years	K – 8	Novice user	Some	VT only
Geri	20+ years	K – 8	Novice user	Some	Both VT & RT
Kim	10 years	K – 8	Some knowledge	Very much	VT only
Margaret	20+ years	K – 8	Novice user	Very much	Both VT & RT
Becca	30 years	K – 8	Novice user	Very much	Both VT & RT
June	20+ years	K – 6	Novice user	Extensive	VT only

their practice. Or as noted by Zhao and Cziko (2001), teachers who, like Gina in Chapter VII, believe they have no need for computers in their classroom activities will not adopt them because they have no obvious advantage over other methods. Through CPI these teachers gained a wealth of knowledge of how computers could support their efforts.[1]

Table 4 gives the reader an overview of the six teachers presented in this chapter. The information provided includes how many years of teaching experience they had prior to participating in the CPI program, what grades were included in their school, how they characterized their prior use of computers/technology prior to the CPI program, and whether they were a Visiting Teacher or also asked to be a Resident Teacher. This information may help the reader in deciding which teacher story is close to theirs.

Mickie

Mickie taught third grade and had taught for 10 years when she participated in CPI in spring 2001. She taught at a school in northwestern Philadelphia where she described the teachers as warm and friendly and the children as really good, until about fourth grade. "Then they [the children] got a bit difficult," Mickie said (anonymous, Personal communication, June 3, 2005). This school was a Kindergarten through eighth grade school. The student population had been declining over the last six years, losing 25 percent of the students in that time. The overall student population was defined as a middle level of enrollments for the purposes of this book.

Mickie said the children came from middle class families and most were bussed into this school from other neighborhoods. The majority (96 percent) of the students were African American, 2 percent Caucasian, and the remaining population divided between Latino/Hispanic and other ethnicities. Average annual attendance was around 93 percent, and only 14 percent student mobility. This made the school population rather stable. About 60 percent of the children who attend this school were eligible for free breakfast or lunch. The principal had been at this school for about six or seven years and provided stable support for technology.

Mickie described herself as computer "illiterate" before she participated in the CPI program. She became involved with the CPI program after she saw a notice hanging in the teacher's lounge for a professional development program that offered free computers and computer training to interested teachers. Mostly she was interested to get the computer since the one she had at the time was old and plagued with problems. She wanted to learn as much as she could about operating the computer and becoming self-sufficient to troubleshoot effectively in order to assist her students. She did not use it at all with her students before taking the CPI training.

I knew that computers were the way of the future. If I was asking my students to do projects using a computer, perhaps this would spur the parents to purchase one for the home. Becoming more familiar with using computers in teaching was a way I believed would benefit my students.

I still don't use computers much. I think the Internet, in particular, can be a real time waster. I remember getting on the computer to do some research once for something and two hours later I was still there. So I don't use the Internet much at all.

The principal is supportive of the teachers. She knows about the statistics on how technology can support both teaching and learning, so she is supportive of the teachers learning about and using computers in their classrooms.

When I took the CPI program, I was in another classroom (in this school). I had a system by which children rotated on and off the computer. They did cooperative projects and writing. I loved it and they loved it. Then I moved to another classroom, and things just fell apart. I could not get the computers to work or could not get onto the Internet, and basically didn't use them that year at all.

In her 2005 classroom, she had three computers, but one did not work. She had them in a corner in her classroom facing the back wall of the room. She had found that when students were working on the computers, those not on the computers were distracted from the lesson if they could see what was happening on the computers. She still had a system for rotating the children on and off the computer, and she had a time keeper to keep track of the time so students did not miss their turn.

Overall, my knowledge of technology definitely increased as a result of the CPI program. Before attending my knowledge level was very low. I only knew very basic computer operations. Now I am comfortable using the software that came with the computer and able to navigate the Web. I'm even using my home computer more. But there is just so much more for me to learn.

My students are quite interested in using the computer for journaling and using the Kidpix software program. The students love the computer and are excited about using it for their own work.

I do believe in the power of computers for student learning, but just find the new core curriculum [of the school district] so demanding that it's hard to find the time. I don't do projects any longer because of the time limit, but I would if I could find the time.

Mickie was not the only person whose classroom practices had been affected by the district's new literacy and mathematics core curricula. Both curricula mandated between 60 and 90 minutes a day dedicated to mathematics and reading, respectively. Even seasoned teachers found it difficult to find ways to infuse technology into these curricula since it was not built into the curricula when developed. Any use of technology was an afterthought.

Geri

Geri taught third grade in spring 2001 when she participated in CPI and had been teaching third grade for five years. When she was interviewed in spring 2005, she was teaching eighth grade for the first time and was the literacy coach for the school. In spring 2001 she was in her 50s, and Geri and the following three third grade teachers were all at the same school. Three of these four teachers participated in CPI in spring 2001 together, and the fourth participated in fall 2001 because of the wonderful experience the other teachers enjoyed through CPI.

This very large school was located in northern Philadelphia. The student population had increased by 40 percent between 1999 and 2006 to a high level, as defined for this book. Sixty-three percent of the students were from low income families and eligible to receive free breakfast and lunch. Student mobility was around 12 percent, which was the lowest percentage of any school presented so far in these chapters. Attendance by students was nearly 93 percent annually. The overall makeup of the neighborhood was low to middle class families, with a school student population of 66 percent African American, 21 percent Latino, 6 percent Caucasian, and 5 percent Asian. While the Latino and Asian populations remained constant between 1999 and 2005, the Caucasian population declined by nearly 65 percent and the African American population rose by 20 percent.

The school completed a new addition in 2002 which added considerable space to the school plant. The library was in the new area and included a computer center with ten computers students could use to work on projects. In addition, there was a state of the art computer lab in the new building used mostly by the Mentally Gifted program. Because the principal was interested in computers and knew the older school did not have the infrastructure for multiple connections per classroom, he bought a mobile, laptop computer lab. The mobile lab was managed by the Mentally Gifted technology teacher and assigned to teachers when they wanted to use the 15 laptop computers on projects with their classes. The mobile lab provided a small computer lab which could be moved into a classroom, rather than the teacher having to schedule the computer lab for a special session. The one mobile lab was so successful, and the school was growing by leaps and bounds, that the principal bought two more to be shared across the entire school. Geri noted that one was not enough to go around to all nine grade levels (anonymous, personal communication, April 6, 2005).

Geri learned about CPI from a notice posted by the principal in the main office. Initially the principal picked five teachers to participate, and then seven more participated the next year. When asked about why she wanted to participate in CPI, Geri noted, "All of us wanted a working computer for our classrooms. We didn't have new computers in our classrooms back then, and we wanted to learn more [about technology]. We had a couple of teachers in the school who were really techies, but most of us were novices, for sure. But, I wanted to learn more about computers" (anonymous, personal communication, April 6, 2005).

As for what she learned from CPI, Geri noted,

We were also given training so we could use it [the computer] with the children, like slide shows, and different things we needed to be taught because we didn't know. We simply didn't know anything about using the computer for ourselves, much less with the children.

The mentors were welcoming and really helpful. They showed us everything. Watching the children was very amazing. They were like well oiled machines. They knew when it was their turn on the computer, when they were to change. It was amazing. There were kids assigned as support for the computers, to fix problems if they arose. They were amazing.

I was glad we were exposed to software we could use for ourselves, like spreadsheets to create a grade book. I thought that was great. I walked away with knowing how to create a spreadsheet, since I didn't know.

I am able to go to the computer and work with the kids on the computer. We did research on the computer. We go to the Internet with the computer in class. Yes, my knowledge of technology absolutely has increased.

One example – I teach low readers and we did a project on whales. We went on the computer and were able to print out information on whales. They did a journal entry on how they would feel if they were a whale. I think there is a real difference for the students accessing information through technology rather than from a book.

As Geri said, she had been teaching for over 20 years, and computers were not part of her teacher training nor of her classroom. But the times were changing and she needed to be part of the new century. She learned how to use the computer for her own needs through using a spreadsheet for a grade book and creating slideshows to use with her students. She also learned how to get her students to easily and equally have access to the three computers she had in her classroom in 2005. With 33 students, it was a challenge to get all the students on the computer, but it was manageable given the examples she gained through her visitation with the Resident Teacher she was assigned through CPI. The original five teachers had the same lunch period so they would share experiences and ideas while they ate lunch, as was true with Debra from Chapter VII and Marta from Chapter V at their school.

When asked how the children worked with the substitute teacher who took their classroom while they were with the Resident Teacher, Geri said that the students were very supportive. The principal knew what was happening, so someone would visit the classroom of the teacher two or three times a day just to make sure everything was all right and running smoothly.

In all, Geri's classroom practices did not change dramatically, but they did change. Before she participated in CPI, technology and computers were nonexistent in her classroom. In 2005, she had three computers and the children used them every week for research and writing, in particular. Third grade students still illustrated stories and drew pictures for science, but typically not with the computer drawing programs. Finding time for such creative practices began to be difficult as the curriculum became more demanding and high stakes testing came into the third grade classrooms in Pennsylvania.

Kim

Kim taught third grade and had only taught third grade at the same school as Geri. She participated in CPI in fall 2001 after hearing what the teachers the previous spring gained from the program. She was in her mid-20s at that time and had only been teaching for a few years. She took a computer class in college but had not used any of those skills with her students. In her classroom, like Geri, she did not have a computer before CPI, so the opportunity to gain at least one computer for her classroom was a real boon.

Those teachers who had been out of college for a while saw CPI as an opportunity to gain skills and materials to help them with their students. Those who were younger, like Kim, and had computing experience in college, were not skilled in using computers with the students. Kim explained it this way (anonymous, personal communication, April 6, 2005):

What I learned was not so much the software and Internet use, but how to use the computers with my children. It was more a class about how to use computers, but we, or those my age, had been using computers since high school. Learning about how to cycle children on and off the computers and actually visiting Resident Teachers to see how they had their classroom run was much more helpful. There were kids assigned as support for the computers, to fix problems if they arose. They were amazing.

Kim further indicated that using the computer as a center and observing other teachers using computers with centers was the best experience of the CPI program for her.

Since CPI, Kim indicated that the principal gave them access to digital cameras. They all learned how to use the camera and to print out images. The principal had been very supportive of these teachers' quest to learn about computers in the classroom. Kim noted that when a teacher was to visit a Resident Teacher, the principal would do whatever was needed to make it happen. He would even send someone from the principal's office to check up on a classroom with a substitute to make sure things were going okay while the teacher was away for a couple of days. While other teachers noted support from their principal for their participation in CPI, this type of support was not reported in any other school.

Overall, Kim's classroom practice did not change significantly. She was already personally comfortable using computers, making slide shows, creating materials to use with her students, and accessing the Internet for her needs. What was most important for her was learning how to use the computer with her children. While she had used computers since high school and took a computer course in her teacher training program, no where along the way had she observed a teacher successfully using technology with his/her children in the classroom, and not in a computer lab. Learning the ins and outs of cycling children on and off the one computer option was significant new knowledge for her, and proved to her that having only one computer in your classroom does not mean the children cannot take advantage of it. This will be echoed by June later in this chapter.

Margaret

Margaret taught third grade at the same school as Geri and Kim when she participated in CPI in spring 2001. In spring 2005, when she was interviewed, she was still teaching third grade. In fact, she had only taught third grade at this school. She was in her 40s in 2001 and had been teaching for nearly 20 years.

When asked how she was chosen to participate in CPI, Margaret said (anonymous, personal communication, April 6, 2005), "It was presented at a staff meeting. The principal asked that anyone interested should see the assistant principal, so I went in and put in a request. I was notified a few days later that I was accepted for it." She noted that she had always felt a little inadequate about computers because she had not had one in her classroom, other than some really old ones in the back of the room. The really old ones were only good to be used as toys to play games, but nothing you could really work on. "I wanted to become computer literate for myself, as well as for my students."

What changed in Margaret's classroom? Before CPI, the students went to the computer lab weekly for 45 minutes. Other than playing games for students who had been "good" in class, there was no other computer use in her classroom. Then, CPI came along. Margaret noted,

We all wanted the materials, the computers and software, and I wanted to become more knowledgeable about computers. I really felt like a fish out of water in regards to technology. I didn't grow up with computers like my kids. My kids can do anything with computers. But I wasn't at that time.

Since I took the program, I have tried to include some simple things with being limited to the one [CPI computer] and a second computer, which we cannot print off of. We have always done journal writing, so what I do now is, every day, I pick one or two students at a time to come up to the computer center after they have written their journal entries and type them on the computer to be filed into their personal folders on the desktop. They really look forward to doing that.

I've tried doing some graphic organizers with Inspiration. I made a math slide show for multiplication tables. I use that to try to help the children in reviewing facts. So far that has worked pretty well.

I remember Dr. Seuss day, when they do something for the day. Last year I had the laptops in my classroom set up for the students to access various sites around the world about Dr. Seuss. We have three carts of laptop computers to use throughout the school. Last year I found sites about Native Americans so, when I was teaching the unit on Native Americans, I used those sites to help the students learn better. The students just loved it. I would never have even dreamed of doing those things before taking CPI!

Did my knowledge increase? I would think so, definitely. I'm definitely more comfortable. I'm not afraid of computers any more. I've always thought that if I pushed the wrong button, the whole thing was going to go up on me. I'm more comfortable with using it and with what's available as far as what you can use it for.

I know the children love it when they get a chance to use the computer. They don't get distracted. They really enjoy being on it. I was doing a graphic organizer on endangered species where they would have to go to the continents and find endangered species. They loved it! My problem was that I only have Inspiration software on the newer computer, so I'm limited to using only one computer at a time for this exercise. It's a bit more time-consuming, but they really enjoy it, the ones who get on. I have yet to get through the whole class on that assignment. I figure next year, some of the ones that I want to do, even if it takes months to do, we will work on it. It's important to the children.

As for my own teaching style, well, I guess it has changed in that I am thinking of something that I can supplement what I am normally doing with something that they can go to the computer to go along with it. I would love to be able to do slide shows with PowerPoint. I did some, but I'd like to do more.

As with the other third grade teachers in this school, Margaret's classroom practices changed, but not wholesale or in a major way. Through CPI she learned how to use graphic organizer software and used it to support her curriculum. She could not have done that before CPI. She learned how to create slide shows and started using them to teach mathematics, something she never would have imagined before CPI. Before she got the new computer that could access the Internet, Margaret's lone classroom computer was an old, antiquated machine that was used only for math games and word processing. Now she could have her students use the one new computer for accessing information on the Internet and both computers to support the literacy curriculum. And, because she was not afraid of technology any longer and was secure in her students' abilities, she brought the mobile laptop lab into her classroom on regular occasions for them to work in groups doing research and projects on those computers as well. None of this was possible before CPI because Margaret was afraid she would destroy the machines, or they would destroy her.

Another outcome, from so many teachers participating in the same program, was that the teachers who took the program at the same time were always comparing notes—how they were doing, what had they tried. "We e-mail things that we create. We share materials we create for math and other subjects. We would never have done that before, but now we help each other out all the time" (anonymous, personal communication, April 6, 2005).

Becca

Becca taught third grade at the same very large elementary school as Margaret, Geri, and Kim. She participated in CPI in spring 2001. When she was interviewed for this chapter in 2005, she was still teaching third grade. One important fact about this school and these teachers was that the staff had been stable for these five years and the teachers still supported one another with the new mathematics and literacy core curricula established in 2002 across the district. As with Margaret, Becca had only taught third grade. She was in her 50s and had been teaching third grade for over 30 years.

When asked why she wanted to participate in the CPI program, Becca said, "We wanted the computers. And because I was truly a novice at the computer and I just wanted to learn how to use it in the classroom" (anonymous, personal communication, April 6, 2005).

Elaine [who taught seventh grade] knew [how to use computers], but we didn't. We would sit with Elaine, because she knew how to do these things. The younger ones, like Kim, had experiences in college. And now with the new building every room has a computer hooked up. I went from having one ancient computer to having six [new computers] in my classroom.

There was a lot of support from the administration for teachers to participate. More teachers wanted to participate than could at any time. Every one wanted to get into the program. The current principal was the same principal then. He was supportive of our visiting other schools.

Learning the Internet and starting journal writing with the children on the computer was most useful for me. And they took [digital] pictures. We learned how children could add [digital] pictures to papers. We created our own homepage. It's probably still up online somewhere, but I don't remember how to access it.

I definitely know more about computing now. I went in knowing basically how to turn the computer on, that's about it. And now I'm not proficient, by far, but I feel much more confident to use them. We help each other. That's a great thing about the teachers at this school – we help each other out!

I have one student in my room who we call the "expert" because he is really good. His knowledge is actually better than mine because I go to him when I have trouble. He goes and sits with other students and they work on the journal writing. I have the students on the computer every other day, if not every day working on their writing.

The kids love to use the computers. At lunch time, we have our expert in the room, and he's allowed to pick two students to work with on the computers. Basically, that's a treat. They love doing the journal writing [on the computer]. Now we print out the journal entries and post them in the hallways for others to see how great my students' writing has become.

Thanks to CPI these four, teachers had more computers in their classrooms and in the school. They started using e-mail with each other. The principal gave them access to digital cameras since they learned how to use them in the program. Becca said, "Field trips, class projects. Taking pictures of these activities. The kids loved it!" (anonymous, personal communication, April 6, 2005).

These teachers had had access to the mobile laptop computer lab for one or two years before they participated in the CPI program in 2001, but they reported they did not know how to use them with their students. With the CPI experience watching children in classrooms with three or more computers, they could envision how these laptops could open up their classrooms to the world. As noted at the beginning of this chapter, Weizenbaum (1976) suggested, long ago, that what man cannot imagine he cannot do. These teachers had access to an incredible set of resources but had no idea or image of how these mobile computer labs could benefit their classrooms. Through their experiences with the CPI program and observing Resident Teachers in their classrooms, the world of computer imagination was opened up for them. While they learned new skills and obtained a computer for their classroom, they also gained confidence to use resources they already had.

In addition, these four teachers, along with the other eight teachers from this school who participated in CPI, supported one another. They shared resources and ideas, and provided moral support along the journey from non-user to user of computers. This approach to peer coaching and mentoring, as described by Berman and McLaughlin (1978), Ball (1996), Fullan and Pomfret (1977), Sharan and Hertz-Lazarowitz (1980), Little (1989), Wilson and Berne (1999), Guskey (2000), and Joyce and Showers (2002), was very powerful for these teachers, as was true for teachers presented in Chapters V, VI , and VII.

June

June taught at a Kindergarten through sixth grade elementary school in northern Philadelphia. This school had lost over one third of its student population between 1999 and 2006 and was classified at the middle level of enrollments as defined for this book. The neighborhood was described as inner city/urban with extremely low income families. Nearly 90 percent of the children who attended this school were classified as eligible for free breakfast and lunch under Title I, which was 17 percent higher than the average for the district. The student population was approximately 80 percent African American, 19 percent Latino, and the remaining 1 percent divided between Caucasian, Asian and other. Approximately 25 percent of these students transferred in or out of the school at some point in the school year, making the school population quite transient in comparison to other schools in the district.[2]

June taught third grade and had a large classroom on the second floor. Before participating in CPI she did not have a computer in her classroom, which was the norm in this school. Her students went to the computer lab once a week for 45 minutes where they learned how to word process and to use various creative programs, but there was no connection between what they did in the lab and what happened in the classroom.

June participated in CPI in fall 2001 with five other teachers from her school. She was extremely nervous about taking the program and was basically there because one of the other teachers saw CPI as a way for them to get computers in their classrooms. June put it this way (anonymous, personal communication, November 6, 2004):

I was so nervous. I thought I was going to break that laptop from the moment they gave it to me. Every time I touched it, something went wrong. Gabby was always right there with me telling me I could do this, that it was going to be fine, that I was not stupid—even though I felt really stupid. She was young and already had a Web site of her own. Why she was there was beyond me, except it was a way for us to get technology for the school.

I really felt dumb! But I got over it. I did practice the lessons and each session [on Saturday] I got better and more confident in my ability to work with that machine. I finally realized that it wouldn't break easily, and I could undo any mistake quickly and easily. I just needed someone to give me encouragement and be patient with me. I'd been teaching for over 20 years and was pretty set in my ways. This was all new to me. But I knew we were going to have to put the report card grades on the computer, so I'd better get over this fear.

When her classroom was visited, June was very proud of what she had accomplished. Not only was she getting the report card grades in on time, but she was requiring the students to cycle through using the two computers she had in her classroom. The children were doing a social studies project on the states. Each child was assigned one of the states to research such topics as the state flag, state motto, various plants and industries prevalent in the state, and the population, among many other facts and figures. She was trying to add mathematics into the social studies by having the children do graphs depicting ethnic make-up of the states. The children would go on the computer and access either Yahooligans or Ask Jeeves for Kids to get the information. One thing June learned through CPI was to establish boundaries for how her children accessed the Internet to keep them safe. They did not print out the information, but would copy the information into a journal. If they found an image they wanted to use, they could save that to their personal folders on the desktop for use in their writing. Then they took that information and made posters that they first presented to each other (e.g., student peer teaching) (Peer Systems Consulting Group, 2006) and then posted either on the classroom walls or outside in the hall way. June remarked, "I would never have done this but for CPI! The examples, the homework, the other teachers all helped me get over my fears. I decided I could do this, and I did!"

June was an exception among the other teachers in her school who took CPI. The one teacher who was her champion, and talked June into participating in CPI, never used the computer she received through CPI with her children, noting that they would just break it. Another teacher had three computers in her classroom which remained unused most of the time, again because she was afraid the children would break them or access inappropriate materials on the Internet. But not June—she changed totally from a fearful, non-user of computers who was afraid of allowing her students to use computers without the computer teacher, to an individual confident in her own knowledge who allowed her students to do research on the computer. When she ran into trouble with software or a Web site, she had learned that it was okay to ask for help.

Summary

The three of the third grade teachers from the same school, Geri, Margaret, and Becca, were also Resident Teachers. They said they wanted to be Resident Teachers to get more training, gain more technology for their classrooms and school, and to help other teachers. Becca noted that "the Visiting Teachers shared ideas with us. I learned from them too. A lot of them were younger and had more experience than we did" (anonymous, personal communication, April 6, 2005). While it may seem odd that these teachers, who described themselves as novice teachers before CPI and not necessarily proficient after CPI, would be asked to be Resident Teachers, these teachers learned to manage their classrooms efficiently such that their students could use the computers they had to effectively support curricular goals. And in the end, that was the point. It is important to remember that CPI was not about teaching reading, mathematics, social studies or science, or supporting any particular grade level or type of student. CPI was about being able to manage the classroom so students could gain access to technology available in the classroom to complete meaningful projects that were curriculum based. Having children type journal entries into the computer to be printed out is a very low level use of computers, and for some (Cuban, 2001) such use would be considered a total waste of the technology given what it can be used for. However, for these teachers, using computers for low-level activities was far ahead of where they were before they participated in CPI. All of their activities and uses of computers need to be put into that context. Change theories acknowledge, as well as the ACOT studies, that there are steps that teachers must go through on their way to using technological innovations with their students (Apple Computers, 1995; Fullan, 1982, 1991, 2001; Hord, Rutherford, Huling-Austin, & Hall, 1987)

Overall, these teachers indicated they learned a lot from the program and still use many of the skills four to five years later. Now they need new computers since the ones they received from CPI are four or five years old. Only one of these teachers indicated that the new district core curriculum was a problem for continuing to infuse technology into her classroom, but she also seemed to be less enthusiastic about how technology could support her students and her own work as a teacher in general. This teacher was from a school where there were 10 other teachers who took the CPI program, but she also reported having problems with her computers when she moved into different classrooms over time. This was also a school where the infrastructure was a very significant problem for all the teachers who wanted to use technology in the classroom.

Analysis

Once again, change does not happen in a vacuum. Each of these third grade teachers had a different personal reason to seek change in classroom practice. Clearly, they had been influenced by the tacit expectation that teachers should use computers in their classrooms, but they did not know how to accomplish that goal. CPI offered (1) to give them a new, working computer, (2) to teach them how to use a computer, (3) to demonstrate classroom management methods for using scarce computer resources with 30 children, and (4) to support them as they took these concepts back to their classrooms. For Mickie, the program changed her practice until the infrastructure of her school introduced barriers, but she still used the computers and her rotation system in spite of problems. As we explore higher grades, barriers to using computers with students in the regular classroom will come into the description. The new 2002 literacy and mathematics curricula established problems for many teachers which will be discussed as they are presented.

Looking at these six teachers, through the theoretical lenses presented in Chapter II, offers insight into how change occurs in these third grade classrooms. These were, for the most part, teachers who described themselves as novice users of computers. Teachers need reasons for changing practice that make the change(s) seem obvious and necessary. The teacher who made the greatest change in her approach toward computers was June. Here was a teacher who was simply afraid of computers and yet readily adopted the use of computers for her third grade students.

Using Rogers' diffusion of innovations theory (1962, 1972, 1983, 1995, & 2003) as a lens, all six of these teachers started at different points in the stages of adoption. Kim had moved through knowledge and persuasion before she became a teacher. She was very comfortable with computers for her personal use, but she wanted to implement computers in her classroom for use with her students and just needed to know how.

June came into the CPI program without any knowledge about computers and not actually wanting the knowledge. She was perfectly happy with her classroom as it was, but she knew she would need to be able to enter the report card grades online eventually. Her reason to participate was not an internal reason but one imposed by the district. However, the CPI program took June from not seeing a need beyond entering report card grades to someone who implemented technology in her classroom on a regular basis. The transformation of June from someone afraid of the computer for her own use, much less that of her students, to someone embracing technology as part of her classroom practice is close to miraculous. But change she did.

Berman and McLaughlin (1978) suggested that mutual adaptation was the best strategy for ensuring change in classroom practices. As with the second grade teachers described in Chapter VII, the needs of these teachers were quite different. Classrooms and teachers each changed as a result of the CPI program. The computers did not change, other than incorporation of newer technology in the classrooms. These were predominantly teachers who simply did not use computers as part of their routine practice, other than perhaps to find materials for themselves or for e-mail. Even Kim, who had taken a computer course as part of her teacher education program and had been using computers since high school, did not use the computer she had with her students. All in all, the one person who changed the least was Mickie, perhaps not by choice but by happenstance. Her classroom changed over time such that she was unable to access resources she had gained and learned to trust through CPI. While she still had students using computers, that usage was very limited. She was the only third grade teacher who discussed the core curriculum as a limiting force on her ability to be creative in the classroom.

Berman and McLaughlin (1978) also discussed effective strategies for ensuring innovations will be adopted in classrooms. These strategies were concrete training, classroom assistance, observation of other classrooms, regular project meetings, teacher participation in decisions, local materials, and principal participation (p. 34). All of these, except local materials, were evident with these teachers. Those schools where numerous teachers participated in CPI afforded these teachers a built-in support network. These teachers often met routinely to share ideas and experiences, which could be equated with regular project meetings. In addition, these peer mentoring/coaching activities supported those who were perhaps more fearful in the beginning that they would not succeed. Having an encouraging principal who understood what the CPI program could bring into his/her school was immensely important for these teachers to feel supported, which appears to have continued over time.

One last lens for analyzing these stories is the stages of teacher adoption of technology proposed by the Apple Classrooms of Tomorrow (ACOT) project (Apple Computers, 1995, Fisher, Dwyer, & Yocam, 1996, Gordon, 1996). June was at the entry level, just beginning to use computers, but moved rapidly past the second level (adoption) to the third level (adaptation) in using technology in her teaching. Margaret moved even further from not using computers with her students at all to using the mobile computer lab with her children, thus changing the classroom teaching and learning environment to support the curriculum in ways she could never imagine before the CPI program, even though the mobile labs were previously available in her school. Kim wanted to use computers with her students but, as noted above, had no mental image of what that would be like. Her experiences visiting the Resident Teachers, as well as networking with other CPI teachers in the fall of 2001, gave her vivid ideas that she could implement in her classroom. Thus, Kim moved from the adoption stage of use into the adaptation stage, giving her students access to a resource for learning they did not have before.

It was only Mickie who did not continue to fully utilize technology in ways she acknowledged learning through CPI. Her circumstances were quite different from the other third grade teachers. Her school was one that was quite old and the infrastructure challenging for the use of computers in the classrooms. These were problems that impacted her ability to use the few computers she had in her classroom when she moved into a classroom with limited support for technology. Mickie never was quite the complete convert to using technology, but she was trying to accommodate her students' using computers to support their work.

Overall, the changes in practice expressed and experienced by these teachers is not what some technology critics would consider radical. To a great extent they support the research findings that the predominant use of computers in elementary schools is for word processing (or journaling for these teachers) and research (or Internet searches). For each of these teachers, moving toward use of word processing and having students do research on the Internet was a radical change from their previous classroom practices. Margaret would never have brought the mobile computer lab into her classroom to use for whole and small group activities if not for her experiences through CPI. For teachers who were unaccustomed and unaware of what computers can bring into the teaching and learning experiences, using computers could be more of a chore than an advantage (Zhao & Cziko, 2001). For teachers who dare to brave the new world of computers, even if that world has been around for nearly 30 years, the revelation can be enormous for them. Not knowing how computers can open up opportunities for learning and expressing or demonstrating knowledge can be very limiting. For some teachers, ignorance is bliss, but not for these six third grade teachers.

References

Apple Computer, Inc. (1995). *Changing the conversation about teaching, learning, and technology: A report on 10 years of ACOT research.* Retrieved on June 6, 2004 from http://www.apple.com/education/K12/leadership/acot/pdf/10yr.pdf

Ball, D. L. (1996). Teacher learning and the mathematics reforms: What we think we know and what we need to learn. *Phi Delta Kappan, 77*(7), 500-508.

Becca, personal communication, April 6, 2005.

Berman, P., & McLaughlin, M.W. (1978). *Federal programs supporting educational change, Volume VIII: Implementing and sustaining innovations*. Santa Monica, CA: Rand.

Cuban, L. (2001). *Oversold and underused: Computers in the classroom.* Cambridge, MA: Harvard University Press.

Fisher, C., Dwyer, D. C., & Yocam, K. (Eds.) (1996). *Educational and technology: Reflections on computing in classrooms.* San Francisco: Jossey-Bass Publishers and Apple Press.

Fullan, M. (1982). *The meaning of educational change*. New York: Teachers College Press.

Fullan, M. (1991). *The new meaning of educational change*. New York: Teachers College Press.

Fullan, M. (2001). *The new meaning of educational change.* Third Edition. New York: Teachers College Press.

Fullan, M., & Pomfret, A. (1977). Research on curriculum and instruction implementation. *Review of Educational Research, 47*(1), 335-397.

Gardner, H. (1999). *Intelligence reframed: Multiple intelligences for the 21st century*. New York: Basic Books.

Gordon, A. R. (1996). *Computers in the classroom: How teachers and students are using technology to transform learning*. San Francisco: Jossey-Bass Publishers and Apple Press.

Guskey, T. R. (2000). *Evaluating professional development*. Thousand Oaks, CA: Corwin Press.

Hord, S. M., Rutherford, W. L., Huling-Austin, L., & Hall, G. E. (1987). *Taking charge of change.* Alexandria, VA: Association for Supervision and Curriculum Development.

Joyce, B., & Showers, B. (2002). *Student achievement through staff development. Third edition*. Alexandria, VA: ASCD.

Little, J. W. (1989). District policy choices and teachers professional development opportunities. *Educational Evaluation and Policy Analysis, 11*(2), 165-179.

Peer Systems Consulting Group. (2006). *Peer Resources.* Retrieved on June 22, 2006 from http://www.peer.ca/profile.html

Rogers, E. M. (1962, 1972, 1983, 1995, & 2003). *Diffusion of innovations*. Glencoe, IL: The Free Press.

Schofield, J. W. (1995). *Computers and classroom culture*. New York: Cambridge University Press.

Sharan, S., & Hertz-Lazarowitz, R. (1980). A group investigation method of cooperative learning in the classroom. In S. Sharon, P. Hare, C. Webb, & R. Hertz-Lawarowitz (Eds), *Cooperation in education,* (pp. 14-46). Provo, UT: Brigham Young University Press.

Wilson, S. M., & Berne, J. (1999). Teacher learning and the acquisition of professional knowledge: An examination of research on contemporary professional development. In A. Iran-Nejad, & P.D. Pearson (Eds), *Review of Research in Education*, Volume 24, pp. 173-209.

Vygotsky, L. S. (1978). *Mind in society: The development of higher mental processes*. Cambridge, MA: Harvard University Press.

Weizenbaum, J. (1976). *Computer power and human reason: From judgment to calculation*. San Francisco: W.H. Freeman and Company.

Zhao, Y., & Cziko, G. A. (2001). Teacher adoption of technology: A perceptual control theory perspective. *Journal of Technology and Teacher Education, 9*(1), 5-30.

Endnotes

1 This book does not suggest that all teachers who participated in the CPI professional development program changed their practices significantly, or at all. In relating June's story, you will note at least two teachers from her school who clearly were not moved to change their teaching practices. For one of these teachers, the visitation to another classroom was not successful. According to the other teacher, her students were incapable of using a computer without destroying it, so she only used her CPI computer for her own work and did not have her students use computers at all.

2 The student population in Philadelphia has been described by some teachers as "migrant." The urban legend is that when report cards are about to be distributed, if children's marks are low, parents simply move to another school thinking that will help their children do better in school.

Chapter IX

Ideas to Concepts: Fourth Grade

Everyone has a technology story. The stories tell about your own levels of adoption and the journey through them. You might not think you use technology because it has been a true tool of your work, the memory of the technology might fade to the background of the project, as it should. If you begin to recall the machines and tool you've used, you will probably surprise yourself (Carroll & Witherspoon, 2002, p.1).

Fourth grade is a wonderful grade. The students are even smarter and more eager to engage in science and social studies activities. Assessment of learning and skill level for reading are more sophisticated in fourth grade. The emphasis is not so much on phonics as much as on comprehension and vocabulary, which are the main themes of literacy through high school.

Like third grade, fourth graders are developing their own styles of learning and engaging with new knowledge. They can discuss and understand difficult subjects. For instance, one fourth grade classroom the author observed for three years discussed "suffrage" in some detail. The student population was mixed, with about one-third African American, one-third Asian and one-third Caucasian, and about two-thirds of the class female. They debated the notion of who was allowed to vote when the American Constitution and Bill of Rights were written back in the late 18th century and what criteria were used to determine such eligibility. This discussion was happening during a presidential election and the teacher was connecting these discussions to current practice where any citizen could register to vote. The students were encouraged to interview relatives at home about their beliefs and experiences with voting and a rousing debate ensued during the next social studies lesson. The point being that fourth graders are old enough to begin discussing political issues and really understand power relationships.

This chapter presents six fourth grade teachers. Three of them are from the same school and participated as a group in the CPI program. The other three are from very diverse backgrounds and schools. Each of them presents a different set of reasons and motivations for participating in CPI, as well as different outcomes for the use of computers in their classrooms.

It is important to point out that, at the time when the CPI program was offered, the School District of Philadelphia required students in fourth and eighth grades, in 2000 through 2002, to complete a service learning project. At the same time the CPI program was offered through the district, another professional development program for infusing computers into project based learning was also offered. Many of the fourth grade teachers participated in both programs since they supported different aspects of the curriculum at that time. In both programs the teachers received an Apple computer for their classroom upon completion; but only the CPI program offered the loaner laptop during the program for them to practice the skills they were learning and to network with other teachers, allowed teachers to visit other teachers' classrooms to see master teachers at work, and provided ongoing support back in the classroom. Assessment of the project based learning program demonstrated that teachers participated to "get" the computer, but they did not actually incorporate the computers into the required service learning projects (Schifter, 2006).

Table 5. Fourth grade teacher summary

Name	Yrs teaching	Grades in school	Prior computer usage	Overall change	CPI teacher
Darrell	5 years	K – 8	Novice user	Some	VT only
Edie	15+ years	K – 8	Novice user	Some	VT only
Dee	20+ years	K – 8	Some experience	Some	VT only
Maya	10 years	K – 4	Novice user	Some	Both VT & RT
Emma	15 years	K – 5	Some experience	Very much	Both VT & RT
Shari	10 years	K – 4	Some experience	Very much	VT only

The participation was what Berman and McLaughlin (1978) would have termed "opportunistic" resulting in no change in normal practice in the classroom. With these six CPI teachers, classroom practice did indeed change.

Table 5 gives the reader an overview of the teachers presented in this chapter. The information provided includes how many years of teaching experience they had prior to participating in the CPI program, what grades were included in their school, how they characterized their prior use of computers/technology prior to the CPI program, and whether they were a Visiting Teacher or also asked to be a Resident Teacher. This information may help the reader in deciding which teacher story is close to their own.

Darrell

Darrell participated with CPI in fall 2000. It was his second year teaching, both years at the same school. Teaching was a second career for Darrell, who previously worked in a transportation industry. He described his school as a large Kindergarten to eighth grade school in northern Philadelphia with students from many different ethnic and socio-economic backgrounds, and English as a Second Language (ESL) as a major component of the school's curriculum. He, and the next two fourth grade teachers, taught at the same school as Eva who taught Kindergarten (see Chapter V) and Jo and Gina who taught second grade (see Chapter VII). When asked why he wanted to take the CPI program, Darrel stated (anonymous, personal communication, December 7, 2004):

At that time [in 2000], I was looking for anything to help me in the classroom. Getting a computer for my classroom and becoming more computer literate was a draw for me. In my previous career I used computers, but for administrative and routine tasks, not for teaching. I wanted to learn more about computers for the administrative purposes of teaching, as well as teaching.

At that time, the staff was more close-knit. All three of the fourth grade teachers participated in CPI together. We had the same lunch period then, so we collaborated on a number of things, including CPI. The principal was beginning to push for more use of computing in the classroom. In 2000, there were a number of teachers who were averse to using computers. One example of the principal encouraging use of computers was that we were to e-mail daily lesson plans to the principal rather than send paper copies. Teachers were not required to submit lesson plans via e-mail, but there were incentives to do so. Some of the teachers were concerned about being required to do something they didn't want to do, and if they complied with one request, they would be expected to do so with the next. I thought using the computer to create lesson plans was so much easier, since you could easily change things around. However, you need to take time to learn a computer program, but if you put the time in at the front end to learn the software, you reap the rewards later on."

Darrell wanted to learn the basics of computing, to be able to navigate and use an Apple computer since he had only used IBM-type computers in his previous work. He learned how to integrate the computer into the classroom through CPI.

For instance, I learned how to have students use the computer for journaling. I still have the journals on the computer. Also I learned how to have the students do projects using the Internet for research. I learned how to use Inspiration for word webs as a whole class exercise. The students loved it!

My main concern when using the Internet with my class was the possibility of students accessing inappropriate material. I also thought Internet searches could be time consuming, but with more practice, training and experience, they were able to use the pre-set Web-portals I set up.

Through one Web site I found two Internet projects that I thought would be interesting for my fourth grade class. "Community Connections" was a project in which students e-mail information about their communities around the country and the globe. This electronic pen-pal project made it possible for students to learn about other states and countries from children their own age. The other one was "The Day I Was Born" (Campanella, 2000), which guides students to links to collect information about what was happening in the news, literature, music, television, and more on the day they were born. Students could use their data to write a report. I also found the "World Weather Watch" (World Meteorological Organization, 2004) site. Every Wednesday the class posted local temperatures, sky conditions, and other meteorological information to a Web site that had data collected by schools all over the world. Students could discuss weather with partner classes from as far away as Australia via e-mail. I thought these types of projects helped my students expand their horizons beyond their local community.

The largest impact on my practice was the administrative aspect. I was new at teaching in general, and to using the computer to support my lesson plans. One of the best things about CPI was learning how to integrate computers into your classroom no matter what kind of computer set up you had – one computer or six, and how to manage 32 children with limited resources.

What also helped was having Resident Teachers right there in this school. What made CPI successful was the realism of being able to see someone else demonstrate in a classroom what you were learning, and having people in the building with you with whom you could share experiences. Yes, having two or three Resident Teachers in my school was great, but being able to share with other fourth grade teachers was great too. These were people who knew and trusted each other, and we could share. This was a major influence for me!"

Darrell shared that he had three computers in his classroom after participating in CPI. He took a couple of other computer courses and received another computer shortly there after. Unfortunately, the school needed to establish a dedicated computer lab for two special reading programs in 2004 and they took away from classrooms computers that had not been earned through professional development activities like CPI. In fall 2004, Darrell only had two computers in his classroom, of which only one accessed the Internet, making it difficult for him to continue what he had learned through CPI. He was mostly using the computer connected to the Internet for posting grades, e-mail, and searching for resources for his teaching, and the other computer was available for student journaling, which ended up being too limited a resource for 33 students.

Edie and Dee

Edie and Dee both taught fourth grade at the same school as Darrell and were interviewed together. All three of them participated in CPI together in fall 2000, and they had the same lunch period when they participated in the CPI program. In fall 2004 the three fourth grade teachers no longer had the same lunch period. Edie and Dee supported each other's teaching as grade partners in this school, so their classrooms were in sync in the curriculum, meaning they were on the same lesson at the same time each day. They shared resources and materials with each other and helped each other out. Their classrooms were right next to each other, and they shared some students as well. Dee had taught at this school for two years before she participated in CPI in fall 2000, but she had been teaching for ten years total. Edie had been at the school for 13 years and had taught for a total of 20 years when she participated in CPI.

Like Darrell, they described their school as very diverse and multicultural. According to Edie and Dee, at least 47 dialects were spoken at this school (anonymous, personal communication, December 7, 2004). There were dedicated ESL groups in each grade level to support these students. Students were from lower economic households, with many considered to be "at risk" children. Ethnically, the population was mostly African American, with Hispanic, Middle Eastern, and Asian students, as well. There were a significant number of immigrants within the student population, hence the large ESL emphasis described by Darrell above. As described in Chapter V (Eve, Kindergarten teacher) and again in Chapter VII (Jo and Gina, second grade teachers), the neighborhood around the school was middle class with row houses across the street and around the neighborhood. Many of these students were bussed into the school from areas of the district far away from the school, which can be said for many of the schools in Philadelphia.

When asked why they wanted to participate in the CPI program, both teachers noted getting a computer for their classroom was definitely a draw because they did not have computers in their classrooms at all in 2000.

I had a PC computer at home for over 15 years, and describe myself as 'computer literate.' I used the computer for personal needs, not educational needs (Dee).

On the other hand, I was not a computer user at all. There was pressure from society and the education community for teachers to be computer literate and to use them to enhance the classroom teaching and learning. It was time [to learn how to teach with computers]. With CPI there was time allotted to practice the skills we were learning, specifically certain specific software and creating a Web site, as well as time to observe other teachers at [this school] teaching with technology (Edie).

I learned quite a lot from CPI, including how to be comfortable on an Apple computer. I am still more comfortable using a PC, but I use the Apple for grades and other school administrative responsibilities.

I introduced my students to Ask Jeeves for Kids. I thought they could handle it. One interesting feature was that when the search term was entered, it was included in several different questions, which could give students ideas about how to narrow down a topic. That's an important skill to learn—how to search effectively and efficiently. Before CPI my main concern regarding the use of the Internet with the students was not with the Internet itself, but with how to manage thirty students in a classroom with only two or three computers.

I was aware of the great potential of the countless digitized resources the Web has to offer our students on millions of subjects. At the same time, the Internet presents us with issues such as readability and source credibility. Also there exists the issue of dealing with a mind-boggling number of links that can lead anyone, let alone our students, to get lost on the Web. As teachers we want to make our kids' experience on the Web a pleasant one. Therefore, it is the teachers' responsibility to structure the Web for the students. We need to make the Web a usable and manageable tool so that our kids don't feel frustrated and overpowered. I think I did that through using Ask Jeeves for Kids (Dee).

I learned how to use and to be comfortable with students using a computer. I have students writing stories which I print and post around the room. I print them because the printer is in another part of the building and having students go pick up printed material is time consuming and they miss out on class time. We, meaning both of us, set up fifteen minute blocks of time for students to use the computers. We each had three computers in each of our respective classrooms. The students knew when it was their time to be on the computers and just cycled on and off. They would rejoin a group lesson or whatever was happening in the classroom. The students worked on keyboarding, journal writing, and other editing skills. We would never have thought of that or even tried it if we had not seen other teachers, through CPI, modeling this for us, or seeing students being responsible users even at lower grade levels.

Before CPI, I was concerned that the children would not have the ability to find topic information on the Internet. I wanted the time on the computer used wisely, and easily. I decided after trying out a number of search engines and kid-friendly portals that my children should not use search engines. The portals were more of a way to break down a task and get just the information I wanted them to find. I never knew about Web-portals before CPI (Edie).

Like Darrell, both Edie and Dee had computers taken from their classrooms to be used for dedicated computer labs, which limited Dee to using the one computer she had left for administrative chores. But Edie still encouraged her students to type papers in class and printed them out to celebrate their accomplishments by hanging them on the walls.

These three teachers added to their repertoire of activities for their students thanks to CPI. Having the opportunity to observe other teachers was an important feature for them. Darrell, Edie and Dee had a common lunch period, like Marta (Chapter V) and Debra (Chapter VII). They came together to share ideas and what they learned through CPI. As noted previously, peer coaching or mentoring was a byproduct at schools with multiple teachers who participated in the CPI experience, becoming a positive force for supporting and sustaining change over time. (Ball, 1996; Berman & McLaughlin, 1975; Fullan & Pomfret, 1977; Guskey, 2000; Joyce & Showers, 2002; Little, 1989; Sharan & Hertz-Lazarowitz, 1980; Wilson & Berne, 1999)

In the end, the curricular needs for literacy in the school and district preempted effective use of computers in the classrooms of these three fourth grade teachers. This did not mean they were not using computers with their students, but the use was more limited when they only had one computer per classroom. Unlike Jo and Gina, who had been Resident Teachers and received large screen televisions to use with computers, these teachers only had the one or two computers left in their classrooms. And, unfortunately, it was easier to relegate the use of limited resources to the needs of the teacher.

Maya

Maya participated in CPI in fall 2000 with Darrell, Edie, and Dee. She was in her late forties and had been a classroom teacher for four or five years at that time. Becoming a teacher was a late in life accomplishment of which she was quite proud. The school where she taught was a small urban school in southern Philadelphia with a mixed population of African American students (75 percent) and Asians students (22 percent). Asian students were Cambodian, Vietnamese, Indonesian, and a small percentage of others. The school had grades Kindergarten through fourth grade with a student population classified as at the middle level, as defined for this book. She described the teachers as very dedicated. Many arrived early or stayed late getting materials ready to hang up in the room or outside in the hallways. A lot of student work was displayed in the hallways to celebrate the students' work. In fall 2000 there was a high element of parent involvement through a parent center. Maya said (anonymous, personal communication, November 8, 2004):

I signed up for CPI because of my experience as a student teacher. I did my student teaching at this school and noticed that every classroom had a large screen monitor connected to a computer. Teachers I observed, and my master teacher for my student teaching, would put the CD-ROM into the computer and display images to the class through the monitor. I really liked how this worked and wanted to be able to do that in my classroom. When I got my own classroom with a large-screen TV, I was having my students copy their homework assignments off the screen. Also I was using the Internet to get images and materials to share to the whole group. I would have the class gather around the monitor. To expand my use and knowledge of the Apple system, I wanted to take the CPI class. I learned about CPI through another professional development program where the CPI program was being discussed. I was one of the few teachers who used the monitor and computer a lot [at that time], so my principal suggested I take the program.

I got an iMac for my classroom and a laptop to do my lessons on. Having the laptop I got an idea about class trip pictures. The class went on a trip and I took pictures of the students. I took the pictures to a [drugstore] to be developed and asked for them to be posted on the Internet. I had never done that before, but thought it might be good. I then took the pictures into a slide show, as a requirement for CPI, and for the children to see. While I might have looked into having the slides posted digitally some day, CPI made me venture out.

From CPI, I tried things I never would have done before. One of the assignments was to develop a home page. I really worked on that assignment until I got it just right, one I really liked. When there was an open house, I had my home page up on the monitor for parents to see what I was doing in the classroom. That was four years ago, and my home page was still up on the Web in 2004. My brother happened to do a Google search for my name, which resulted in my home page coming up. I was amazed that it was still there, but there nonetheless. I had not updated it since I put it up for CPI.

I keep anecdotal notes about my classroom and children on the computer like a daily log. When I have parent conferences or need to take information to a meeting, I print it off the computer. This has been so successful.

Unfortunately, I am not using the computer as much now that I have a new computer with [the new Mac operating system]. The operating system is cumbersome and it's not as easy to use with the TV. This is not about CPI, but about the computer. I like my older computers better.

One project she described doing with her students in 2002 was developing a book. She would gather the students around the large screen TV and they would brainstorm topics for writing a book. They would generate about four or five ideas and she would ask the class to vote for their favorite topic. After the topic was determined, they would list ideas about the topic to include in the book. For instance, if they were writing about a particular state or country, they would include information about the population, the language, food, agriculture, politics, and customs, among other topics. She would then post, on the board, the list of topics for students to volunteer to research and write about. They would also illustrate their writing using one of the illustration programs she had, like KidPix or a paint and draw program. Once the children had completed their writing, they would edit and print out their work to post around the room. They would also save the work into a folder on the computer for the book. Maya had a contact with someone who would then take the computer files and create a bound book for the class. The book would have the names of all the children who contributed to the book and a dedication to parents in the back. Each child received a copy of the book to give to their mother. Maya said it was not uncommon for parents to ask for additional copies to give to a grandmother or aunt.

The book idea, as described by Maya, required knowing someone who could actually publish a book. However, Parker (see Chapter VI, First Grade) "published" student work himself and this worked for his classroom. The point of the project was to printout (publish) student work, not that it needs to be bound like a book. There are other ways to printout and "bind" student work like a book.

Like the previous fourth grade teachers in this chapter, Maya reported being hampered by the demands on classroom time by the new literacy and mathematics core curricula. While she did not experience having computers removed from her classroom for new literacy programs, her discomfort was with the tight curricula leaving no room for individualization, and a new computer with a new operating system she did not understand. Even so, her classroom was full of materials her students had created through technology, and she continued to encourage her students to use technology to demonstrate what they have learned.

Emma

Emma taught at a Kindergarten through sixth grade elementary school in northwestern Philadelphia. Sixth grade was new to the school in 2006, and the school would add seventh and eighth grades in the next few years. She had taught there for nine years, but for a total of 15 years overall. She taught in a fourth grade classroom with 25 children, more boys than girls. As Emma put it, "each class has its own personality and personalities. On average they are well behaved and of average intelligence" (anonymous, personal communication, June 1, 2006). She had a few students in the past who had been accepted to an academic magnet school in the city and she was very proud of those students. In 2006 she had some students reading on the ninth grade level and some who were reading far below grade level challenging her ability to diversify instruction. In all, her class was probably a better than average class for fourth graders in Philadelphia.

This school had a student population classified as at the middle level, as defined for this book, and which had decreased approximately 15 percent since 1999. There was a temporary building for adding sixth grade classes in 2006. Attendance was at 94 percent, which was very high for the district, and mobility of the students was approximately 14 percent in a given year. The neighborhood around the school was single family homes and characterized as a middle class neighborhood.

Emma described her teaching style as structured and teacher directed. (anonymous, personal communication, June 1, 2006)

I have the students do a lot of independent work to encourage them to own their own knowledge, following instructions and following through on assignments. When they [the students] ask for explicit instructions, I remind them that I have given them the instructions and they should read them and try to figure out any problems they have first before asking questions.

I try to have the students do interesting projects, not boring ones. I want them to be self-regulated and to do independent work. Not all work is collaboration. But I also want the work to be connected to the real world. While the students may not have the knowledge yet to be completely successful, the object is for them to have the confidence to try.

In spring 2006, she had six computers along the back wall of her classroom, plus one computer that she used for her own work. She had recently received a laptop as part of the district-wide program. Five of the six computers accessed the Internet. The seventh computer would become part of her computer center the next year giving more opportunities for her students to do research on the Internet and practice their computing skills. The building became a wireless environment in 2006, allowing her to access the Internet from the new laptop.

The best part was visiting other classrooms. That was so cool! The chance to see peers teaching, sharing materials, bringing back materials was great. I'm always looking around for how to take new ideas into my classroom. I came back from those visits with so many ideas, and not just about using computers in the classroom. One idea I came back with was identifying 'computer experts' from my students. These would be the students who would help the other students on the computers if they got stuck or were having a problem. I routinely had one or two per class to take on this responsibility.

And CPI was all about collaboration. Even the Saturday sessions were an opportunity to talk with other teachers about what they did in their classrooms, and, what I called 'forced time to play' with computers. Can you imagine? Forced time to play on a computer? And the program director was incredible.

Emma had a number of software programs she encouraged her students to use to support the curriculum. She set up a schedule, just as she learned through both the CPI Saturday sessions and the teacher she visited. They used First in Math to hone math skills, along with one of the Carmen San Diego® series software, Reading Blaster,™ and I Spy™ programs. She also had an off-line keyboarding program the students used to learn typing skills. Emma said, "I tell them that when they go to college, if they have typing skills, they can charge other students to type their papers for them. It's a way to make money. They need to see a reward for their efforts, and money talks."

One year she did a multidisciplinary project with her students on pets. They used AppleWorks to make pamphlets and brochures about different pets. She also had a template for a project on national monuments. In each project, the students did research on the Internet, brought that information into AppleWorks, and created something that demonstrated their new knowledge. Emma said she was looking for more such projects and spent time looking for resources on the Internet, as she learned through CPI.

Emma was pursuing a master's degree in education through an online opportunity in Pennsylvania. She said the students in that program talked about different topics and the discussions were interesting since these were teachers from across the state, not just in Philadelphia. It was another opportunity to share with others and learn from others—exactly what CPI was all about.

Sheri

Sheri had taught fourth grade in the School District of Philadelphia for 10 years, all at an elementary school in western Philadelphia. In fact Sheri noted that she did her student teaching at that school and was asked to stay. The school had grades Kindergarten through fourth grade and had a student enrollment classified as low, as defined for this book. The school was a small, but caring, environment. According to Sheri, the principal there was very supportive of the teachers and students and had established an environment of excellence (anonymous, personal communication, June 14, 2006).

Sheri did have a computer in her classroom before she participated in CPI. And after CPI she participated in another professional development program and received another computer for her classroom. In spring 2006 she had five computers in her classroom set up with two chairs per computer so students could work collaboratively on projects. She noted:

All classrooms at [this school] have multiple computers and the principal has established a culture of using technology to support teaching and learning. The school recently received smart boards for each classroom and they are being used extensively. Since the smart board in my classroom is connected to the Internet, I use the smart board to set the stage for what students will be expected to do when they use the computers that day.

The CPI training was described by Sheri as taking a person from where they came into CPI, or from what level of technology knowledge and comfort existed at the beginning, and helping each individual to become more comfortable and knowledgeable. She came into the program relatively comfortable, at what she indicated as a mid-level of computer knowledge. While the instructors were working with other teachers who were just learning how to turn on the computer, which way to hold the mouse, or how to open an application, she would work with a neighbor in the lab who was more proficient in using computers and try things on her own.

Two specific aspects of the program were very important for Sheri. First was the process of establishing lesson plans using computers that plugged in the curriculum standards for mathematics, literacy, social studies, science, health, and all the other performance standards. And second was creating a Web page.

The Web page became something that I used continuously until it disappeared during the spring of 2006. I would have the students doing entries on the Web page regularly. I kept the families of my students aware of things there and created mailing labels with the URL suggesting they visit the site to learn what was happening in her classroom. When I sent something home for a parent to review, I put a label on it encouraging them to visit the Web site. I changed my Web page regularly and added links to pages with resources for my children to use. When the computers [in her classroom] came on, they opened to my Web page. Visitors could write to me on the guest page. I trained the children to go to the appropriate sites for research. I had links to Baltimore Aquarium and other Web sites that related to the curriculum.

This was a resource that made sense to Sheri and she used it extensively. And as she noted, "The children are more and more experienced with the Internet, so it is imperative that I bring these resources to bear on the curriculum."

Her classroom was set up such that immediately on the right hand wall, as you came into the classroom, was the computer center, with five computers, headsets, and maps of the world all over. The desks were in the middle of the room, and about half of the room was the rug area for group teaching. The smart board was there and she used it to access the Internet frequently to get ideas for the computer sessions. As Sheri noted, there was so much in the core curricula now that it was challenging to incorporate technology, but she tried to incorporate as much as possible. Word processing and Web searching were two activities that were easy to include as part of lessons. Getting her Web page up and functioning again would be very important as Sheri reached out to parents and her students outside of school, which she thought was best done through the Internet.

Summary

All of these fourth grade teachers wanted to expand their own repertoire of teaching skills and opportunities to help their students learn. They saw CPI as a way to gain a computer for their classroom, and also a way to gain skills to support their own teaching. As with the third grade teachers, the district-wide core curricula and literacy initiatives appeared to have reduced opportunities for using the computer skills they developed through CPI to the three most common uses of computers by teachers: basically administrative uses, word processing, and Internet searches. However, each of these teachers indicated that they understood the importance of infusing computing into classroom activities and did as much they could, given the constraints. Edie, Maya, and Sheri still printed out student work, posting them around the classroom to celebrate their accomplishments. Having only one working computer in a classroom made it difficult for Edie and Dee to give every student a quality opportunity to express their knowledge through writing, multimedia or other options.

Why was it that, beginning in fourth grade, each of the teachers indicated that the core curricula had curtailed their use of computers with their students? Why was this not as pervasive in earlier grades? This is difficult to ascertain from the interviews. If high-stakes tests began to be implemented in fourth grade, then that could be a reason; however, in the Commonwealth of Pennsylvania, those tests begin in third grade. Starting in 2002, the district began to require students in grades 3 through 10 to take the TerraNova/Supera, 2nd Edition,[1] national, norm-referenced standardized tests that assess achievement in reading, language, math, and science. Test results were used as one source of information for assessing potential student performance on the high-stakes tests, but those results were not available until the beginning of the next school year. Thus, fourth grade is the first chance for teachers to begin to target students at risk for performing poorly on high-stakes testing required by No Child Left Behind legislation (Department of Education, 2001). Starting in the fall 2005, the TerraNova/Supera tests were given in October, with results available earlier for determining students needing extra help in preparation for the high stakes tests given in March.

Another possible reason for these teachers expressing less time in the day for including activities using technology, other than word processing and Internet searches, could be the core curricula in the district, as noted by several of these teachers. The core curricula were established by content experts using national standards for literacy, mathematics, social studies and science, and did not infuse technology resources or ideas for teaching the lessons in non-traditional ways into the curriculum. Not that this was an excuse for teachers to not explore other options, but it was reported that the curricula were very tightly scheduled. The core curricula were introduced in 2002 and only addressed literacy and mathematics at that time, as required by NCLB (U.S. Department of Education, 2001). Following the directions of NCLB, core curricula for science was introduced in fall 2005, followed by core curricula in social studies. With the addition of each standardized core curricula, teachers perceived that their freedom to explore ways of teaching their particular students was being limited. And with limited opportunities for innovation in the curriculum, teachers were aware of what they absolutely had to do in the classroom, and anything extra could be minimized.

Analysis

Did change happen for these fourth grade teachers? Certainly, there was change. Technology critics would argue that the change was not sufficient (Cuban, 2001), but the circumstances in some of these schools precluded their desire to use technology in meaningful ways to support the curriculum.

Using the Perceptual Control Theory, proposed by Zhao and Cziko (2001), these were teachers who understood the need to seek options for their practice at educating students. These teachers were using their personal political capital to influence what happened in their classrooms (Huebner, 1999). None of these six teachers were required to learn about computers for classrooms. Edie and Dee alluded to the fact that there were both societal and professional pressures, or expectations, for teachers in the 21st century to understand and use computers to support teaching and learning, but their principal was not forcing them to participate in the CPI professional development program. On the contrary, they understood that the program was a mechanism for personal and professional growth, as well as an opportunity to gain technology resources for their classrooms. Little did they know that eventually those resources would be the only computer resources in their classrooms.

Emily and Sheri, on the other hand, welcomed the opportunity to learn from colleagues. In fact the CPI experience, for them, was all about collaboration with other teachers. Being able to visit another teacher, to observe another teacher's classroom, and to see concepts about how to infuse technology into classroom practice in action were all critical components for them learning how to change their classroom practices.

Using change theory proposed by Fullan and Pomfret (1977) and later Fullan (1982, 2001), changes occurred in each of these teachers classrooms. Roles changed for each of these teachers. No longer were they the sole source of information for their students, other than textbooks. Students were using the computers to find information on their own, supplementing what the teachers could provide and the textbooks contained. However, these changes were not sustained due to factors outside the control of the teachers. In one school, computers were removed to support a literacy program mandated by the district. In another school the fourth grade teacher's computer system was changed. In spite of these detrimental changes to their options, all six of these teachers continued to value the principles and practices they learned through the CPI program. They continued to value the need to ensure that all students use computer resources in meaningful ways that support the pedagogical goals of the curriculum, whether through projects, writing journal entries to improve their writing skills, using the Internet as a resource for research given scarce resources in classrooms in many schools, and developing skills that would serve the students well as they grow into adults.

Darrell had only been teaching for two years before he took the CPI program, and Maya had only been teaching for four or five years. They had not truly established routines for their classrooms, and indicated they were searching for ideas to expand their options. Emma noted that she continuously, for over 15 years, sought new ideas to engage her students in meaningful classroom work around the curriculum. Bringing computers into the mix of experiences, indeed, expanded their range of pedagogical options.

From the Concerns Based Adoption Model (Hord, Rutherford, Huling-Austin, & Hall, 1987), each of these teachers came into and graduated from CPI at different stages of adoption. Darrell, Edie, and Dee initially moved from the awareness that they needed to learn more about Apple computers and how to infuse technology into their classrooms, through the management stage where they were concerned about managing time, to the consequence stage where they were concerned about the impact of these computers on their student's learning. Unfortunately for them, they regressed back to either the management stage (Edie) or personal stage (Darrell and Dee) due to the fact that the school removed computer resources from their classrooms. At the same time, the level of use for these three teachers initially moved from non-use (level 0) through preparation (level 2) and mechanical use (level 3) to routine use (level 4A) with their students. One could argue that they remained at level 4A because they continued to routinely use the computers. However, Darrell and Dee only used the computers for administrative requirements, like entering grades for report cards and attendance records; but Edie continued to have her students use the computers for journaling and papers, which she printed out to post around the classroom.

Maya and Emma also moved toward the consequence stage of adoption with the various projects they developed for their students to complete. The book created by Maya's students gave them opportunities for exploring a topic, expressing their new knowledge in essays, illustrating their essays in creative ways, and demonstrating to Maya and their parents their learning and skills. Similarly, Emma's students' project on pets allowed them to express their new-found knowledge in ways that were less cumbersome prior to having access to computers. Editing essays before word processing was a very difficult and time consuming process. Anyone who remembers typewriters before correct-a-tape will identify with the editing process before digital word processing. They too reached the routine use of technology level (level 4A), but did not go further to individualized instruction.

Maya and Emma also reached the collaboration stage of adoption. Both talked about sharing ideas with other teachers and how important that was for them. Maya, in particular, spoke about the need to reach out to other teachers and not be so isolated in her classroom. Sarason (2002) wrote about the isolation of teachers in their classrooms and the lack of communication even across grade levels in schools. But Maya did not want to be so isolated and saw CPI as a mechanism for gaining access to other teachers to learn from them and to share with them what she had learned. Similarly, Emma relished in the collaborative aspects of CPI and continued to collaborate with other teachers through her online master's degree program.

The ACOT studies offered five levels of adoption of computers into classroom practice. (Fisher, Dwyer, & Yocam, 1996; Gordon, 1996) Each of these fourth grade teachers entered the CPI program passed the entry, or first, level of adoption and the adoption, or second, level as well. All of them had used computers before, to some extent, but were not proficient with using these tools with their students. Each moved into the adaptation, or third, level of adoption where they were using their new computers to engage their students in ways different from before. None appeared to have reached the appropriation stage where they were really changing their roles in the classroom, creating individually paced instruction or using more complex projects.

Rogers theory of diffusion of innovations had five stages of adoption as well. His theory was not specific to computers in classrooms, as was the ACOT findings, but quite applicable (1962, 1972, 1983, 1995, & 2003). Each of these six teachers had moved through the initial knowledge stage in that they were aware of computers and aware that there was an expectation from the education world, real or imagined, that teachers should be incorporating computers into classroom practices. From this, they did not actually need to be persuaded, because they believed, at least for Edie and Dee, that it was time for them to come into the 21st century. All made a conscious decision to participate in the CPI program so they could learn more about how computers could support their teaching and students' learning. Whether they reached the final confirmation stage is questionable. Initially, all six integrated computers into their classrooms. When interviewed only two were using computers at any level similar to that demonstrated immediately after participating in CPI. Darrell, Edie, Dee, and Maya were all influenced by changes occurring within the district. Darrell, Edie, and Dee had computers removed from their classrooms to facilitate implementing new reading programs for the school. Maya's practice was changed when the operating system for her computer was upgraded but she did not find it as easy to use as the older version. Only Emily and Sheri were extensively using computers in their classroom. Sheri was, perhaps, in the best position with the new smart board.

The support of the principals of these six teachers was not as evident as in past chapters. This is not to say the principals in these schools were not supportive. In fact, Darrell, Edie, and Dee all taught in the same school as Ellen (see Chapter V), Jo, and Gina (see Chapter VII). This principal was extremely supportive of the teachers' participation in the CPI program and provided new technologies for them to use as well. Unfortunately, that same principal was required, due to scarce financial resources, to muster computer resources to meet the demands of new literacy programs. Sheri specifically mentioned how supportive her principal was in the use of computers in the classroom, including the addition of the smart boards in all classrooms. This small school was truly blessed.

If classroom teachers are focused on the development of the child, then these teachers reached out to resources to help them with this process. Maya, Emily, and Sheri specifically spoke to wanting, not needing, multiple ways to engage their students in meaningful activities for learning. Computers have been presented by many sources as one way to improve student learning. (Liao & Bright, 1991; Lipsey & Wilson, 1993; Schacter, 2001; Sivin-Kachala, 1998; Wenglinsky, 1998) Were these teachers, like the others described in this book, merely following the crowd? Perhaps not, given that they were, for the most part, engaged in the CPI process of allowing computers—as a learning tool—to empower their students to learn in creative ways, as proposed by Jerome Bruner (1960), Lev Vygotsky (1978), and John Dewey (1938).

References

Ball, D. L. (1996). Teacher learning and the mathematics reforms: What we think we know and what we need to learn. *Phi Delta Kappan, 77*(7), 500-508.

Berman, P., & McLaughlin, M. W. (1978). *Federal programs supporting educational change, Volume VIII: Implementing and sustaining innovations*. Santa Monica, CA: Rand.

Bruner, J. S. (1960). *The process of education*. Cambridge, MA: Harvard University Press.

Campanella, C. (2000). *The day I was born*. Retrieved on June 22, 2006 from http://www.stphilipneri.org/teacher/dayiwasborn/

Carroll, J. A., & Witherspoon, T. L. (2002). *Linking technology and curriculum: Integrating the ISTE NETS standards into teaching and learning*. Second edition. Upper Saddle River, NJ: Pearson Education.

Cuban, L. (2001). *Oversold and underused: Computers in the classroom*. Cambridge, MA: Harvard University Press.

Dewey, J. (1938). *Experience and education*. New York: Macmillan.

Fisher, C., Dwyer, D. C., & Yocam, K. (Eds.). (1996). *Education and technology: Reflections on computing in classrooms*. San Francisco: Jossey-Bass Publishers and Apple Press.

Fullan, M. (1982). *The meaning of educational change*. New York: Teachers College Press.

Fullan, M. (2001). *The new meaning of educational change*. Third Edition. New York: Teachers College Press.

Fullan, M., & Pomfret, A. (1977). Research on curriculum and instruction implementation. *Review of Educational Research, 47*(1), 335-397.

Gordon, A. R. (1996). *Computers in the classroom: How teachers and students are using technology to transform learning*. San Francisco: Jossey-Bass Publishers and Apple Press.

Guskey, T. R. (2000). *Evaluating professional development*. Thousand Oaks, CA: Corwin Press.

Hord, S. M., Rutherford, W. L., Huling-Austin, L., & Hall, G. E. (1987). *Taking charge of change*. Alexandria, VA: ASCD.

Huebner, D. E., (1999). Politics and the curriculum. In Hillis, V. (Ed.) (1999). *The lure of the transcendent: Collected essays by Dwayne E. Huebner*. Mahwah, NJ: Lawrence Erlbaum. (pp. 15-22). Originally published in Passow, A. (Ed.) (1962). *Curriculum crossroads*. New York: Teachers College Press, pp. 87-95.

Joyce, B., & Showers, B. (2002). *Student achievement through staff development*. Third edition. Alexandria, VA: ASCD.

Liao, Y. K., & Bright, G. W. (1991). Effects of computer programming on cognitive outcomes. A meta-analysis. *Journal of Research on Computing in Education, 24*, 367-380.

Lipsey, N. W., & Wilson, D. B. (1993). *Practical meta-analysis*. Thousand Oaks, CA: Sage.

Little, J. W. (1989). District policy choices and teachers professional development opportunities. *Educational Evaluation and Policy Analysis, 11*(2), 165-179.

Rogers, E. M. (1962, 1972, 1983, 1995, & 2003). *Diffusion of innovations*. Glencoe, IL: The Free Press.

Sarason, S., Edited by R. L. Fried. (2002). *The skeptical visionary: A Seymour Sarason Education Reader*. Philadelphia: Temple University Press.

Schacter, J. (2001). *The impact of education technology on student achievement. What the most current research has to say*. Santa Monica, CA: Milken Exchange on Education Technology.

Schifter, C. C. (2006, March). *Technology in teaching: Comparison of three staff development programs.* Paper presented at the annual meeting of the Society for Information Technology in Teacher Education, Orlando, FL.

Sharan, S., & Hertz-Lazarowitz, R. (1980). A group investigation method of cooperative learning in the classroom. In S. Sharon, P. Hare, C. Webb, & R. Hertz-Lawarowitz (Eds), *Cooperation in education,* (pp. 14-46). Provo, UT: Brigham Young University Press.

Sivin-Kachala, J. (1998). *Report on the effectiveness of technology in schools, 1990-1997*. Washington, DC: Software Publisher's Association.

U.S. Department of Education. (2001). *No Child Left Behind Act*. Washington, D.C.

Vygotsky, L. S. (1978). *Mind in society: The development of higher mental processes*. Cambridge, MA: Harvard University Press.

Wenglinsky, H. (1998). *Does it compute? The relationship between educational technology and student achievement in mathematics*. Princeton, NJ: Educational Testing Service Policy Information Center.

Wilson, S. M., & Berne, J. (1999). Teacher learning and the acquisition of professional knowledge: An examination of research on contemporary professional development. In A. Iran-Nejad, & P.D. Pearson, (Eds), *Review of Research in Education*, Volume 24, pp. 173-209.

World Meteorological Organization. (2004). World Weather Watch. Retrieved on June 22, 2006 from http://www.wmo.ch/web/www/www.html

Zhao, Y., & Cziko, G. A. (2001). Teacher adoption of technology: A perceptual control theory perspective. *Journal of Technology and Teacher Education, 9*(1), 5-30.

Endnote

1 The TerraNova/SUPERA, 2nd Edition are nationally-normed standardized tests that assess student achievement in reading, language, math, and science. The Supera is the Spanish version of the TerraNova/Supera and is administered to students in Spanish bilingual programs. The district first administered the TerraNova/Supera in the fall of 2002. The fall 2002 test was administered to students in grades 3-10. In spring 2003, and continuing through spring 2005, students in grades 1-10 were administered the TerraNova/Supera.

Chapter X

Expressing Knowledge in New Ways: Fifth Grade

Using a computer once meant having to write a computer program to make it work. Today's microcomputers, however, are very user-friendly. This is due partly to changes in computers and partly to improvements in software. Operating a classroom microcomputer today is easy primarily because software publishers have produced lots of ready-to-use computer programs for classroom applications (Geisert & Futrell, 2000, p. 5).

How do fifth grade classrooms differ from fourth grade classrooms? The simple answer is that the curricula for mathematics, literacy, science, and social studies are more complex, requiring the students to learn new vocabulary as well as to expand on concepts previously learned. The room will not actually look much different, other than there may be less student work posted around and more rubrics posted for writing, reading genres, ciphering mathematics, or science concepts. Word walls will no doubt still be prevalent, since the vocabulary word of the day is still an important component of learning. And there may be many different word walls, such as one for mathematics or one for science concepts, as examples. Otherwise, the room itself will not appear much different from a fourth grade classroom or sixth grade classroom.

How different are the students other than at least one year older? In Philadelphia, fifth grade is when some students may be accepted into academic magnet schools, thus siphoning away from neighborhood schools those students who achieve at high levels. This results in students in typical classrooms being those who may be more academically challenged. At the same time, these children are typically 11 years old, or pre-teens. So there is another dynamic coming into the classroom that was perhaps not there in earlier grades. Thus teachers must have teaching skills and a repertoire of teaching methods that engage these students in subjects in which perhaps they have not excelled nor been interested in the past—not an easy situation to navigate.

In Philadelphia, the fifth grade may be located in an elementary school or middle school. The fifth graders may be the older students in the school with seniority over the younger students; or they may be the younger students in the school and at the low end of the student pecking order. Each situation establishes different dynamics between classes of students. Being the oldest, or even one of the older grades in a school, may give the students opportunities to mentor younger students or to take on leadership roles within the school. However, being the youngest and newest group in a school means having to learn the culture of the school along with new teachers and subjects. Another feature of being in an elementary school setting is that the fifth grade students are assigned one teacher who is responsible for all subjects, except for special subjects like art and computers. If the fifth grade is part of a middle school, students typically have a homeroom teacher with whom they meet at least once a day, but then they rotate through the school to different teachers for English, social studies, mathematics, science, gym, health, art, and/or music. While

Table 6. Fifth grade teacher summary

Name	Yrs teaching	Grades in school	Prior computer usage	Overall change	CPI teacher
C.C.	20+ years	5 – 8	Novice user	Extensive	Both VT & RT
Venda	10 years	K – 8	Some knowledge	Very much	VT only
Edith	30+ years	K – 5	Novice user	Some	VT only
Patti	30+ years	K – 5	Novice user	Some	VT only
Chris	20+ years	K – 6	Novice user	Extensive	Both VT & RT
Gail	30+ years	K – 6	Novice user	Extensive	Both VT & RT
Dionne	20+ years	K – 6	Novice user	Very much	VT only

they have a homeroom teacher, this teacher is not responsible for all subjects giving them a different status to the student. Having such a schedule can be disconcerting, while having one dedicated teacher can be "safe."

This chapter presents seven fifth grade teachers from different areas of the district, all but one from elementary schools. Four teachers are from the same elementary school, and two others are from another elementary school.

Table 6 gives the reader an overview of the teachers presented in this chapter. The information provided includes how many years of teaching experience they had prior to participating in the CPI program, what grades were included in their school, how they characterized their prior use of computers/technology prior to the CPI program, and whether they were a Visiting Teacher or also asked to be a Resident Teacher. This information may help the reader in deciding which teacher story is close to their own.

C.C.

C.C. taught at a middle school in northern Philadelphia. She taught reading and language arts to fifth graders in the morning and science in the afternoon. According to C.C., this middle school was one that the parents from the neighborhood elementary schools petitioned the school district to establish (anonymous, personal communication, March 30, 2005). They wanted a new middle school so their children could stay in the neighborhood, closer to home.

The school had grades five through eight, so the fifth graders were the youngest in the school. The makeup of this school had more Latino students (55 percent) than African American students (34 percent), followed by Asian and Caucasian students in percentage. Nearly 82 percent of the students were classified as coming from low income families making them eligible for free breakfast and lunch. This percentage was well above the district average of 73 percent. The student population had been declining slightly since 1999 but was still classified as a high level of enrollment, as defined for this book. Student attendance had been steady for six years at 89 percent, again below the district average, with student mobility averaging at 15 percent over the same six years.

With the CPI pilot program, C.C. received five computers for her classroom. C.C. described her becoming involved in the early CPI program this way:

The principal asked me if I wanted five computers [for my classroom], and it was really hard to say no. I didn't know what was involved and had very little technology skills. I had a computer at home, but didn't have Internet access. I thought it was time for me to learn.

First, and most important, the kids were really excited about the computers. They were truly motivational tools to help them get through dry stuff. Writing a piece on the computer, creative writing, daily journal assignments, all became easier for them to do on the computer. The 'sloppy copy' kids moved to the computer and all of a sudden you could read their work. Some didn't know about computing before this, but they caught on really fast. It was a tool for those with problems in writing to get over their issues with writing and penmanship.

Thirty-six students with five computers was a challenge. We set up guidelines—sit quietly and do work, then get to work in teams on the computer. One child well-versed [with computers] would do training of other children. During my prep time, I would get one or two students and teach them what I was learning.

Then there was the summer program. The kids who came wanted to be there. They did projects, worked in rotation teams. They learned Hyperstudio®, and multimedia. It was an eye-opener for me. Clearly, computers are a motivational tool for kids.

When I left Philadelphia a few years ago, I moved to New Jersey. I'm now a computer teacher and studying for a master's degree in technology. I teach robotics and Web design to my kids. I would have never thought I would be doing this 10 years ago when I participated in the CPI pilot program. Not in a million [years], but now I'm comfortable with my skills, and definitely not afraid. All because of CPI!

What was unique about C.C.'s story was how she went from being an overall middle school teacher, teaching language arts and science, not familiar with or using computers to support her own work or the learning of her students, to someone not only comfortable with her own computing skills but who taught robotics and Web site design. She credited the CPI program with showing her the power of computers in bringing middle grade children to want to write and to be excited about learning. The transformation was not just about C.C., but also her classroom and the learning opportunities for her students.

C.C. was not the only teacher presented in this book who made significant strides in their own use of computers. Like Marla, a Kindergarten grade teacher (see Chapter V), and Parker, a first grade teacher (see Chapter VI), C.C. moved from being a classroom teacher to being a computer teacher. Perhaps the only other teachers presented so far who made similar strides were Gina, a second grade teacher (see Chapter VII), and June, a third grade teacher (see Chapter VIII), who went from being total non-users of computers and non-believers in the power of technology to supporting their classrooms and exploring and exploiting uses of computers with their students. No, these teachers were not like other teachers, but their stories demonstrate that the CPI program touched teachers in different ways, showing them different approaches to helping their students soar higher in learning with technology.

While C.C. and June did not reach the same level of change in their classroom practices, with June continuing to be a third grade teacher and C.C. moving on to teach in the computer classroom, each made major changes in their perception and personal use of computers to support their students, assigning projects to their students that required the use of the computers to complete. Neither of these teachers would have imagined doing such before participating in the CPI program. Imagination opens up practice to new ideas (Weizenbaum, 1976).

Venda

Venda had taught at her school for seven years. She taught second grade for five years, and had taught fifth grade for the last two years. She really liked fifth grade students. She said you can talk with fifth graders in ways you cannot with second graders. Venda taught at the same school where Mickie taught third grade (see Chapter VIII). She said she participated in CPI in spring 2001 (anonymous, personal communication, June 3, 2005).

Being a teacher is a second career option for me. I majored in business in college and minored in information management. I worked in business for a couple of years and discovered it just was not the environment for me. So I went back and got a teaching certification and here I am loving what I do.

Given my background in information management, which means a lot of computer use and programming, I wanted to participate in CPI because I wanted my students exposed to the computer. I knew programming and databases, but using computers in the curriculum was a different story. There were no computers in my classroom, and most of the children did not have a computer at home. CPI introduced me to how to use the computer with the children, how to manage a classroom with a small number of computers and give children a chance [to use the computer to show what they know and what they could do].

When her classroom was visited, all three computers were in use by students doing research and creating presentations or papers for different purposes. The students clearly knew what was expected of them. They had their turn on the computer to complete a component of their presentation and then they cycled off so the next group of students could work on their projects. They were quite efficient in their use of their computer time, which was something Venda had learned from her CPI Resident Teacher.

Now I use First in Math and PowerPoint with my students. It was difficult with the second graders because they needed a lot of assistance. This is not the case with fifth graders. I get them started with a project and they work on their own to get things done. I credit CPI with helping me feel comfortable allowing children to use the computer independently.

My Mentally Gifted children are doing projects for Career Day. They are creating their projects using PowerPoint and will present them to the school on Career Day. It's a way to empower them to demonstrate their knowledge and ability. I would not be doing these things if it weren't for CPI.

These students were researching different career options by using the Internet to find the vital information they needed. They found images they could use in their PowerPoint presentation, and used the "links" feature within PowerPoint to link out to an Internet site to demonstrate specific features of a career. Truly, they were empowered by Venda's trust in their abilities to be successful.

Venda made it clear that she wanted her students to have as many opportunities as possible to learn in different ways and to express that knowledge in different ways. Through the three computers in her classroom she and her students could accomplish these goals. She came into teaching with a thorough knowledge of the power of computers through her experiences in the business world and her use of programming languages. She knew very well how computers could harness information for people to turn into knowledge to be used, not just a bunch of facts to memorize. She did not know, however, how to transition this power from the business world to the education classroom. Venda credited her experiences with the CPI program and her Resident Teacher connection with helping her understand how to best use the three computers she had in her classroom in spring 2005. CPI provided her with examples for managing her students using computers and resources to support her students' learning. In addition, she gained the confidence to allow her fifth grade students to work independently, knowing that empowering them to be responsible would produce quality learning. She used cooperative learning groups to accomplish her goals, which was not in her teaching repertoire before her CPI experience.

Edith

Edith taught at the same school in northeastern Philadelphia where Ronnie, who teaches Kindergarten (see Chapter V), and Sally, who teaches second grade (see Chapter VII). Edith had taught at this school for six years out of a total of 30 years teaching experience, and she was in her fifties when she took the CPI professional development course. She taught fifth grade and participated in CPI in spring 2001. As noted earlier, this school had 47 percent Caucasian students, approximately 10 percent Hispanic students, and the remaining student body fairly evenly divided between Asian and African American students. Only 45 percent of the students were considered to be from low income families and eligible for free breakfast and/or lunch. The school had very high expectations for achievement, and the students generally rose to the occasion scoring above both city and state averages.

This school was a Kindergarten through fifth grade school, so the fifth graders were the oldest students in the school, unlike the middle school where C.C. taught, where they were the youngest. The population of this school had increased lightly by about 6 percent. Attendance was steady at 93 percent, and mobility was approximately 18 percent. Edith said: (anonymous, personal communication, June 3, 2005)

I was a 'novice' user, with minimal use. We had this one teacher in the school we considered to be the expert, and often sought her out for help.

The number one reason to participate in CPI was to get a computer for the classroom. However, I participated because of peer pressure and the need to know more.

The principal was supportive. I had an incident with the principal shortly after I arrived here. I brought [to him] a note to send to a parent and asked for his approval. While he had no problem with the content of the letter, he said, 'We usually word process letters to go home.' I thought the letter looked fine, but clearly the principal was telling me something.

Edith indicated that the core curricula had slowed things down for using technology, but she used the reading curriculum Web site and Inspiration[1] to support her teaching. She learned about Inspiration through CPI. Projects by students using both options were posted in the hallways demonstrating how successful the students were with using these programs and how successful Edith was with helping her students' learning using concept mapping of ideas to explore stories, novels, and social studies. Again, none of these activities would have been imagined by Edith were it not for her CPI experiences, as imagination is what helps people to create new ways of thinking and doing (Weizenbaum, 1976, p. 18).

Patti

Patti taught at the same school as Edith. She had taught at this school also for six years out of a total of 31 years teaching experience. She also taught fifth grade and participated in CPI in fall 2001. Patti noted that the school expected the students to achieve at high levels, and she indicated that the students scored above average on high stakes-testing (anonymous, personal communication, June 3, 2005).

The school had had a computer lab for many years, but had not always had a computer teacher. The current computer teacher had been there for 11 years. Some teachers had a computer in the classroom, but not all. One advantage for Patti was that the computer lab was right across the hall from her classroom; so if she wanted all her students to participate in a particular software option, she scheduled the lab. She did that to teach her students how to use a software program called Tesselmania®[2] which taught them all about creating tessellations, where students explore geometric shapes with symmetry and more.

I would classify myself as a 'non-user' of technology. It was not something that was important for me, so I didn't use it. So many of the other teachers were participating in CPI and this was a way for me to get a computer for my classroom. I honestly did not expect to change my teaching practices, but somehow the technology brought options to the classroom not there before.

Like Edith, Patti indicated that the school district's core curricula put barriers up precluding the use of technology. She used the reading core curriculum Web site, but also used Tesselmania and Inspiration to support her teaching. Patti noted that it was through CPI that she was willing to explore Tesselmania due to her new found confidence with technology. This particular school had students' projects using both software options posted in the hallways along with other projects, demonstrating how successful the students were. Patti noted that none of these activities would have been imagined, or even tried, before her CPI experiences. As she said, she was a "non-user" of technology, so why should she consider any options other than what the curriculum required?

It is important to note that Edith and Patti did not indicate that they had made major changes in their classroom teaching practices; however, they were willing to explore new options. Both had been teaching for over 30 years, long enough to establish expected ways of teaching for them, and expected ways of learning and demonstrating knowledge for their students. The CPI experience demonstrated to both of them that there were options that could expand their repertoire of pedagogical tools, and also provide opportunities to expand their students' options to empower their own learning.

Chris

Chris had taught fifth grade for over 20 years and had moved recently into some administrative responsibilities in her school. The school where she taught was in northwestern Philadelphia and was the school where Emma taught fourth grade (see Chapter IX). The school was a Kindergarten through sixth grade school. The students were predominantly African American (98 percent), but there are a few Asian, Hispanic, and Caucasian students as well. Chris indicated that 64 percent of the students were classified as coming from low income families and thus eligible for free breakfast and lunch, and that the neighborhood of tree-lined streets was comprised of single-family, middle class homes, with some low-rise apartment buildings (anonymous, personal communication, April 13, 2005).

After a sabbatical to learn more about language and changes happening in schools in general, Chris came back and started to bring computers into her classroom. She reported that the principal wanted this to happen and challenged the teachers to use computers, e-mail, and taught them many skills. For Chris, the principal was a strong influence. His goal was to make their school like any [wealthy] suburban school. So, when CPI came along in spring 2001, he recommended she apply along with another fifth grade teacher. They took the course together, which supplied a support mechanism for both of them.

My Resident Teacher was at a school nearby. I spent a full day looking at computer programs, Hyperstudio projects, and getting ideas for using the computers with my children. I wanted to give children a chance to use computers. I numbered my computers and set up a rotation system. I established a student assignment system to (1) find the weather of the day and share out to the class, and (2) find a 'black fact for the day' and share out to the class. My students' computer and communication skills improved. Prior to CPI, computers were only used for drill and practice. After CPI the children enjoyed using computers more and supported each other. Students did go to computer class for 45 minutes each week, but what they did in that class and what we did in my classroom were not connected in any way.

Next, I moved into having students doing projects and reports on the computers. They would work with a partner to research information and type up their reports. One year they did a project on endangered species. Each pair of children was assigned an animal. They were to respond to specific questions, get pictures from the Internet, and create slideshows. None of this could have been done before I took the CPI program.

Then, I became a Resident Teacher. I created a booklet for the teachers and a slideshow that demonstrated the school and the schedule. I showed the Visiting Teachers how I rotated my students on and off the computers to accomplish their goals, how to use certain CDs, and to create journals. It was different being a Resident Teacher and having teachers visit my classroom. I could share with other teachers, learn what the Visiting Teacher knew, and more. It was very rewarding. If using the computer is important, you will do it.

As proposed by Arlene Kramer and Jerry Snyder in Chapter I, the principal at this school was an important factor in getting his teachers to use computers more in their classroom activities. While he was a major factor in providing the impetus for participation in the CPI program, the program itself provided the evidence for Chris to change her practice in her classroom. She moved from being predominantly a teacher-oriented teacher to a more student-oriented teacher, who allowed the students' experiences and ideas to help them learn (Dewey, 1938, & Vygotsky, 1978).

Gail

Gail had taught elementary school for 34 years. She had taught in the School District of Philadelphia for 30 years, of which 29 of those had been at the same Kindergarten through sixth grade school in northwestern Philadelphia, the same school where Chris taught fifth grade. Most of her teaching had been in fourth and fifth grades, but she had taught every grade except Kindergarten and second grade.

Gail's classroom was an open and airy classroom with large windows across one wall. There were typically lots of student work hanging around the room to celebrate their knowledge and skills. The room was very inviting and comfortable, and the 10 computers in the back of the room were easily accessible and available for students.

Before she participated in CPI, Gail had two very old computers in her classroom. "I knew very little about the programs that were on the computers or how to use them. Unfortunately, I had an Apple computer here at school and an older PC computer at home. So trying to merge the two was a terrible experience. So I signed up for the [CPI] program and learned so much I ended up buying an iMac for home," said Gail (anonymous, personal communication, June 7, 2006).

It gave me such a great interest in learning more. I went from [knowing] nothing [about computers in classrooms] to being able to do so many things. It was such a pleasure to teach the children [using the computers]. When the computer teacher had the children for 45 minutes, it was really only 30 minutes between getting to the computer lab and then getting back to class. Trying to learn the basics, like spelling, spacing between words was a challenge. She never had time to get to those skills because of time constraints. So I could work with them [in the classroom]. They would type their journal entries at least once a day and use the spell checking [feature]. They had their own folders where they kept their work on the computers. And they learned so much.

And then we did slideshows. You could read a lot of things, but once you put it into a slideshow the topics came alive. I would have a computer out here [in the hallway] at report card time for parents. There would be a slideshow on it for them to see what their children had done. I can't rave enough about what I learned through CPI.

She enjoyed the slideshows so much she had the children do a slideshow project each grading period, or three times a year. Then they did two per year. One project she had done with her children for the last five years, since CPI, was to create a cookbook of recipes. It was a variation of a "how to" book report project she had done with her fifth grade students for 15 years. The students went through cookbooks, found a recipe they liked, and talked with whoever cooks in their house about the recipe. They would make the dish together and bring it to school for a luncheon. This recipe would be their contribution to the book. Then they invited teachers and staff in the school to contribute to the cookbook as well, giving them a copy in the end as a "thank you." The children worked together to create the cover page that would be a special tribute to their mothers for Mother's Day. They determined which recipes were starter courses, main courses, salads, and desserts, and created divider pages for each section. The back of the book had the names of all the people who contributed to the book, including who created or found the various graphic images. These books were gifts to the children's mothers on Mother's Day. Gail noted, "They [the mothers] love it and the students love making it. They work in groups. They have to learn all the skills—designing the covers and more, working collaboratively. They just love it."

To continue their writing experiences with the book, the day after the luncheon they had a reflection sheet where they wrote about their experience—what they liked, did not like, and so on. In this part of the project, the students worked on their writing skills to be clear and concise while making sure they explained their answers to the questions. For Gail, this was an extension of the literacy and writing curriculum, one from which the students learned, and one from which the parents saw outcomes they liked.

Every time there was a computer class, Gail signed up. She said she felt like such a novice that she thought she would break something on the computer. If the instructor were there showing her what to do, she was fine, but when she was out of that situation, at home or in her classroom, she felt lost again. Then came CPI where she learned to be self sufficient and developed self confidence in her skills and comfort level, and developed an appetite for learning more about how computers could support her classroom.

You know you really can't teach what you don't know. They [the computers] were back there [in the back of the classroom] and I knew how to get my e-mail, but that was it. That was as far as I was going. The children knew more than I did. And then I was learning things, I could teach them. And if something went wrong, the children would say 'let's try this.' They had more experience than I did.

In spring 2006, Gail had 10 computers in her classroom. All of them were used for word processing, among other activities. Every day the students were on the computer. In her ideal world she would have them doing slideshows. They did research for projects. She gave them her e-mail address and she encouraged the students to e-mail her materials for her to review. Some of her students came in early before school started and some stayed through lunch to work on projects. There was not much time during the scheduled day, but some students took advantage of these additional times to work on the computers. She had set time from 8:30 to 9:15 AM for First in Math and other computer work.

One of our principals, who died a few years ago, he was really into computers. He got a new computer lab full of new computers. He took the older computers and put them into classrooms in groups of computers in the back of the classroom. I got a mini-lab in the back of my classroom. Then I started looking for more options.

She participated in an Accelerated Reader™ program that gave her a computer and $1,000 worth of books and computer disks to go with them. She went to every program that gave away a computer. She said, "The more [computers] you have, the more students can work on them."

It made me a much better teacher. I really do believe that. Now I don't have problems with new technology ideas because I have a background. I can help other teachers because I have confidence in my skills. Not that I'm a computer teacher, but I can help other teachers.

One other outcome from CPI was that the fifth grade teachers at this school participated together, getting together to share when ever they could. As noted with Chris, there were a number of teachers from this school who participated in the CPI professional development program. The advantage of this outcome has been discussed previously, but the advantage here was that there were three fifth grade teachers, including the next teacher discussed, who got together regularly to compare their CPI experiences and to share resources and ideas. This peer coaching or mentoring was an outcome of the CPI program that was not planned, but clearly one that was an advantage in schools where principals encouraged multiple members of his/her teaching staff to participate in CPI to gain technology for their schools, to increase their personal knowledge of using computers, and to gain experiences and ideas to support student learning through technology (Ball, 1996; Berman and McLaughlin, 1978; Fullan and Pomfret, 1977; Guskey, 2000; Joyce & Showers, 2002; Little, 1989; Sharan & Hertz-Lazarowitz, 1980; Wilson & Berne, 1999).

Dionne

Dionne had taught in the School District of Philadelphia for 23 years and at the same Kindergarten through sixth grade elementary school for 20 years. She had always taught fifth grade and was asked to teach sixth grade for one year. She did not like teaching sixth grade, mostly because the other sixth grade teachers were in another part of the school buildings and she felt lonely without their support. When she was a fifth grade teacher she had the camaraderie of the other fifth grade teachers in the same wing of the building. Dionne taught at the same school as Chris and Gail.

Before she participated in the CPI program in 2002, Dionne had one or two older computers in her classroom. She did not use computers for much of anything at that time. Dionne said she used the Internet to get ideas for activities, but not until CPI did she get the students using computers for class work (anonymous, personal communication, June 7, 2006).

Through CPI I created a slideshow on Native Americans, which was a large unit in the fifth grade curriculum. The students were excited about that.

After CPI she only had four computers in her classroom, so she could not have all the 28 students in her classroom using computers at the same time. She set up groups and gave them assignments for research and projects, and the students took it from there. She said, "I told them I was learning about using computers and how it was going to support their learning and how excited I was about that. They seemed to understand that I was doing this for them."

They [the students] did slideshows too, with a lot of my assistance. I would coordinate this with the computer teacher, who wanted to make sure she was supporting classroom activities as well.

As noted above, Dionne said the fifth grade teachers worked together and shared ideas. They shared and collaborated on Web sites and materials. "Getting together with other teachers was really a boon."

Typically Dionne had 26 to 33 students in her classroom. The enrollment had dropped between 2002 and 2006 and the smaller class sizes made it had been easier to get more students using the computers in her classroom. Also, having student teachers every year had helped as well.

After CPI Dionne felt more comfortable with technology and willing to get on and explore. She, like so many of these teachers, changed her classroom practices ever so slightly. While her use of computers with her students may not come up to the standards of some technology critics (e.g., Cuban, 2001), her use of computers with her students supported the curriculum. She chose to accept her principal's challenge to move her teaching practices to another level with computers for extending the writing and literacy curricula, and to provide alternative methods for her students to demonstrate their knowledge.

Summary

Teaching fifth grade is in the middle school grades, even if not located in a middle school, per se. The students need to be challenged, excited about learning, and see the relevance to their lives. They are at the age where doing "creative" projects with many creativity computer programs seems "childish" to them. Getting these students into exploring concepts through concept mapping programs, or presenting research through slide show presentations, puts the power of learning into their hands, thus empowering them through technology. These teachers saw the impact of computing on the students in their Resident Teachers' classrooms and came away believing they could make the same changes in their classrooms. As a testament to these radical ideas, they were having children create presentations for Career Day, posting tessellations created by their students for all to see, creating cookbooks from recipes their mothers gave them, and more. Each of these projects is perfect for this age range that Jean Piaget classified as "concrete operational," which he suggested lasts from ages 7 to 11 (1972, 1990). Fifth graders, if they are on normal track, are 11 years old, and probably moving between the "concrete operational" stage of development and the "formal operations" stage. They still need very concrete examples and activities from which to learn but are just beginning to grasp abstract ideas. Using the Tesselmania software program brought the abstract concept of tessellations to life for Patti's students.

For Chris, becoming a Resident Teacher gave her the opportunity to share with other teachers and to learn from them. For Gail, CPI gave her the confidence to want to know more about computers and to share that knowledge with her students. The new core curricula did have an impact on classroom computer activities. Gail, in particular, noted that she used to have students doing slide show projects at least three times a year, which she whittled down to once a year simply because there was not enough time in the day. In spite of changes in the curriculum and requirements, they seemed to be able to see the power of technology where it fit into the curriculum and made things happen for their students. For all of these teachers the principal was a major supporter of their venture into infusing computers into their classroom activities.

Analysis

Once again Cuban's (1986, 2001) portrayal of teachers having computers in classrooms that languish from lack of use might be seen in these teachers' stories. It is true that they did not use the computers they had every day, all day, but judiciously as needed or as the project required. They had been willing, even if cajoled, to participate in the CPI professional development program because use of computers was expected by their principal, parents of their students, and other educators. Going into the program not fully committed to the process was not a detriment for these fifth grade teachers. They came out of the CPI experiences with fresh ideas for how they could serve their students best. Remember, the teacher with the least experience as a teacher was Venda, for whom teaching was a second career. Computers were second nature for Venda as a tool for her own needs, but learning how to use them productively, effectively and efficiently with her students was important for her. Gail, on the other hand, had taught for 34 years, which afforded her enough time to become really rooted in her ways as a teacher, especially since most of those years were teaching at the fourth or fifth grade level. However, Gail was a savvy teacher who understood that there were changes happening in classrooms nationally, given these things called computers. Gail's experience with the CPI program gave her the confidence and drive to learn more, whetting her appetite to learn as much as she could about how computers could support her classroom, giving her the confidence to become a Resident Teacher herself.

Using Rogers' (1962, 1972, 1983, 1995, & 2003) stages of adoption of an innovation, each of these teachers came into the CPI program wanting to learn how to use computers with their students. They made the decision that this was important for them and their students, and from their experiences, moved to implementation of computer options in their classrooms. CPI gave them the opportunity to understand what computers bring to projects and exercises that cannot be captured through use of traditional chalk board or chart paper, and how the software programs they learned were compatible with the curriculum, even if the curriculum changed after their graduation from CPI. They were afforded the opportunity to try out and learn more about the computer and software options they were to use. And most importantly, these teachers mentioned how vital it was to observe another teacher in her classroom successfully managing 33 students using a limited number of computers. It was all fine for the CPI instructors to explain and describe how to do these things, but only when it was demonstrated in a real, live, functioning classroom that did not explode into chaos did these teachers, along with all the other teachers described in this book, come to understand that they could be as successful in their classrooms.

According to Berman and McLaughlin (1978), the most successful implementation of an innovation in a classroom is through mutual adaptation. These teachers adapted their classroom practices to accommodate students' use of computers to support their learning. And again, as noted by Berman and McLaughlin, effective strategies for change include classroom assistance, opportunities to observe the innovation in other classrooms, project meetings, teacher participation in decisions, local material development, and principal participation. These seven teachers did not speak about classroom assistance, per se. Unlike other teachers who mentioned the support by the CPI program director or the local educational support staff, they were mute on this subject. However, Edith and Patti were at the same school where there were a number of teachers who participated in CPI. They got together regularly to share ideas, even if across grade levels. Chris, Gail, and Dionne were all from the same school as Emma (see Chapter IX). They talked about how the first three teachers who participated in CPI got together regularly to share. These local peer mentoring or coaching opportunities have been described in almost every chapter, and were the local support these teachers needed to move forward. (Ball, 1996; Berman & McLaughlin, 1978; Guskey, 2000; Joyce & Showers, 2002; Little, 1989; Sharan & Hertz-Lazarowitz, 1980; Wilson &Berne, 1999)

Using the Concerns Based Adoption Model (Hord, Rutherford, Huling-Austin, & Hall, 1987) as an analytic lens, each of these seven fifth grade teachers was aware of computers and how to use them. None of these teachers were totally non-users of computers. While Patti described herself as a non-user, this was in the context of the classroom. Each of these teachers moved through the informational and personal stages of concern relatively quickly, wanting to know how to manage their classrooms effectively. While classroom management was of importance, since losing control was deemed a sign of weakness as a teacher, the most important use of computers in classrooms was to support student learning. None of these teachers talked about using the computers they had as merely reward for good behavior. On the contrary, each discussed projects students completed, whether using the Tesselmania software program to understand the mathematics of tessellations, or creating slideshows and presentations using PowerPoint as described by Venda and Gail.

Collaboration was a stage that was very evident for Edith and Patti, along with Gail, Chris, and Dionne. Both groups of teachers were teaching in schools where there were a large number of teachers who participated in the CPI program. Having the opportunity to meet with other CPI teachers during a break, before or after school, or during lunch, was a tremendous advantage over those teachers who were perhaps the only teacher in their school to take advantage of the CPI professional development program. For the four teachers at the same school, each teaching in the same grade level, they were afforded opportunities to share success stories and projects that would be beneficial for other students in the same grade. This sharing might not be so successful if sharing between a fifth grade teacher and a second grade teacher since the students and the curricula are so different.

All of these teachers noted that the core curricula impacted their ability to infuse computer activities into their lessons. Given the nature of the curricula, they were painfully aware of the need to be on lesson each day and week. In the early years of the core curricula, established in 2002, the fifth grade students were the first required to take the Pennsylvania System School Assessment (PSSA), described in Chapter VII. Fifth grade teachers were all aware that their students needed to be prepared to successfully take the PSSA, since this test was used to determine Adequate Yearly Progress for the school as required by the NCLB (U.S. Department of Education, 2001). This outcome meant there was little "wiggle room" in the day to veer from the structure of the curriculum outline, and thus fewer opportunities for teachers to incorporate computer assignments.

Lastly, looking at the stages of development purported through the ACOT studies (Fisher, Dwyer, & Yocam, 1996; Gordon, 1996), these teachers had developed significantly from how they used computers before their CPI experiences. None of these teachers entered the CPI program at the entry level. They all indicated that they had basic knowledge of how to use computers, in general, just not how to use them with their students. One could argue that Edith, Patti, and Dionne developed skills to move into the "adaptation stage" and used their new technology to increase student productivity and engagement. Having said that, they did so in ways they would never have tried before the CPI program, including the willingness to ask the school to invest in a computer program to teach tessellations to their students. In contrast, C.C., Venda, Chris, and Gail used the computers in their classrooms to move their students to higher learning activities. They created cooperative learning groups for projects. In some instances they attempted to create projects that were cross-curricular and required the students to begin thinking in abstract terms. The ACOT studies suggest that teachers need several years to reach the "appropriation" stage of development in infusing computers into classroom activities. This might be true where there were limited opportunities for the ACOT teachers to observe and interact with other teachers learning the same skills. For the CPI teachers, they not only had their observation experiences with the Resident Teacher to whom they were assigned, but also the support of other CPI teachers in their schools.

Conclusion of Section III

These 19 teachers taught in different schools and areas of the city, as did the Kindergarten, first and second grade teachers interviewed. Getting a computer for the classroom was clearly a draw to participate in the CPI professional development program for most of these teachers. At the same time, many of them identified themselves as either "non-users" of computers or "computer illiterates" giving them reason to participate in the program. While they started out as "illiterate," they came away with a sense that they could accomplish great things in their classrooms with their students using computers. They developed systems to help their students make the most of using the computers they had, and, in the case of two teachers, made the most of starting a project in the classroom, polishing it in the computer lab, and presenting it back in the classroom. They combined learning through technology with enhancing communication skills to help their students learn. The key, again, was what Weizenbaum suggested in 1976, that man cannot create that which he cannot imagine. These teachers learned to imagine how their classrooms could be different from the status quo, and how their students could both learn and demonstrate that learning through technology.

References

Ball, D. L. (1996). Teacher learning and the mathematics reforms: What we think we know and what we need to learn. *Phi Delta Kappan, 77*(7), 500-508.

Berman, P., & McLaughlin, M. W. (1978). *Federal programs supporting educational change, Volume VIII: Implementing and sustaining innovations*. Santa Monica, CA: Rand.

Cuban, L. (1986). *Teachers and machines: The classroom user of technology since 1920*. New York: Teachers College Press.

Cuban, L. (2001). *Oversold and underused: Computers in the classroom*. Cambridge, MA: Harvard University Press.

Dewey, J. (1938). *Experience and education*. New York: Macmillan.

Fisher, C., Dwyer, D. C., & Yocam, K. (Eds.) *Education and technology: Reflections on computing in classrooms*. San Francisco: Jossey-Bass Publishers and Apple Press.

Fullan, M., & Pomfret, A. (1977). Research on curriculum and instruction implementation. *Review of Educational Research, 47*(1), 335-397.

Geisert, P. G., & Futrell, M. K. (2000). *Teachers, computers and curriculum: Microcomputers in the classroom.* Third edition. Boston: Allyn and Bacon.

Gordon, A. R. (1996). *Computers in the classroom: How teachers and students are using technology to transform learning*. San Francisco: Jossey-Bass Publishers and Apple Press.

Guskey, T. R. (2000). *Evaluating professional development*. Thousand Oaks, CA: Corwin Press.

Hord, S. M., Rutherford, W. L., Huling-Austin, L., & Hall, G. E. (1987). *Taking charge of change.* Alexandria, VA: Association for Supervision and Curriculum Development.

Joyce, B., & Showers, B. (2002). *Student achievement through staff development*. Third edition. Alexandria, VA: ASCD.

Little, J. W. (1989). District policy choices and teachers professional development opportunities. *Educational Evaluation and Policy Analysis, 11*(2), 165-179.

Piaget, J. (1972). *The psychology of the child.* New York: Basic Books.

Piaget, J. (1990). *The child's conception of the world.* New York: Littlefield Adams.

Rogers, E. M. (1962, 1972, 1983, 1995, & 2003). *Diffusion of innovations*. Glencoe, IL: The Free Press.

Sharan, S., & Hertz-Lazarowitz, R. (1980). A group investigation method of cooperative learning in the classroom. In S. Sharon, P. Hare, C. Webb, & R.

Hertz-Lawarowitz (Eds), *Cooperation in education,* (pp. 14-46). Provo, UT: Brigham Young University Press.

U.S. Department of Education. (2001). *No Child Left Behind Act.* Washington, D.C.

Vygotsky, L. S. (1978). *Mind in society: The development of higher mental processes.* Cambridge, MA: Harvard University Press.

Weizenbaum, J. (1976). *Computer power and human reason: From judgment to calculation.* San Francisco: W.H. Freeman and Company.

Wilson, S. M., & Berne, J. (1999). Teacher learning and the acquisition of professional knowledge: An examination of research on contemporary professional development. In A. Iran-Nejad, & P.D. Pearson (Eds), *Review of Research in Education*, Volume 24, pp. 173-209.

Endnotes

1 Inspiration® is a software program that helps students with concept mapping, among other skills.

2 TesselMania™® is a software program where students can learn about, and how to create, tessellations, which are unique, interlocking, puzzle like designs.

Section IV

Teachers Voices:
Middle Grade Teachers (6-8 Grades)

Middle school is another world from the early childhood grades and elementary grades. The children are teenagers with all the trials and tribulations that go along with being those ages. The students who are academically strong have moved to the academic magnet schools leaving the students who are more academically challenged behind. And then there is the issue of whether the grade is in an elementary school or middle school which brings another set of challenges. In either case, teachers are now more likely to be subject specialists than generalists, which could make it even easier to infuse computers into the curriculum because the teacher is only dealing with one or two subjects. Alas, that is not the case. In fact, as will be presented, infusing technology into middle grade classrooms appears to be much more difficult than in earlier grades.

The classroom activities move toward a combination of more discussion of ideas and exploration of concepts, moving toward higher order thinking and problem solving. The computer activities support the curriculum through expressing concepts through PowerPoint® presentations, using Inspiration® to map concepts, and using the Internet for research for information all to demonstrate knowledge in unique ways. These three chapters present six sixth grade, six seventh grade and seven eighth grade teachers who participated in the CPI program.

As with the teachers from Sections II and III, these teachers were asked to respond to four basic questions listed in Section II Introduction (p. 89). The stories of the nineteen teachers will begin with some background information about the school and context in which they teach to set the scene. Throughout these chapters, the names of the teachers are pseudonyms to protect the identities of these teachers who volunteered to be interviewed for this book.

Chapter XI

Demonstrating Learning with Computing: Sixth Grade

In the early 1980s it was possible to be optimistic about the diffusion of computer use in education because hardware costs were decreasing rapidly and many interesting ideas were in the air; today, we seem to have reached at least a temporary cost plateau that is very high indeed for all but the wealthiest school districts (Scott, Cole, & Engel, 1992, p. 220)

Like fifth grade, sixth grade may be in an elementary school or a middle school. In my elementary school way back when in time, sixth grade was the highest grade. This meant we were the oldest in the school and had leadership opportunities—like helping out in the school office or in the cafeteria. We were the "big kids on campus," so to speak. In some middle schools in Philadelphia, sixth grade has been the youngest group. As noted with fifth grade, there is no doubt a difference in the experiences for students depending on whether they are in an elementary school setting or middle school setting. To such ends, many middle schools are being phased out of the district in exchange for elementary schools that have Kindergarten through eighth grades. Two of the schools represented by these sixth grade teachers were undergoing such changes starting in fall 2005.

In either case, and as noted with fifth graders in Chapter X, the students who have not been admitted to one of the few academic magnet schools with middle grades are the ones for whom learning is more likely a challenge. Teachers in these classrooms are more likely to be challenged to engage these students in subjects they don't understand, perhaps because they have trouble with the subject matter, or perhaps they are reading well below grade level and, therefore, the reading materials are difficult. When reading the material is difficult, students are more likely to be off task or act up.

In any case, sixth graders are becoming more independent people. They are 12 or 13 years of age, meaning they are either pre-teens or teenagers with all the angst and thrill of physical and emotional development that goes along with those ages. While teachers must teach specific subjects like reading, social studies, mathematics, and science, they are challenged also to handle the growth factors that come with pre-teens and teenagers. This is not a topic for this book, but one that can impact teaching and learning opportunities and activities in classrooms.

This chapter presents six sixth grade teachers. All but one taught in a middle school setting, and one taught in an academic magnet school where her students, unlike the other five sixth-grade teachers, were not the oldest in the school, but clearly not the youngest. These issues may become of value in thinking about the different experiences these teachers had through the CPI professional development program.

Table 7 gives the reader an overview of the teachers presented in this chapter. The information provided includes how many years of teaching experience they had prior to participating in the CPI program, what grades were included in their school, how they characterized their prior use of computers/technology prior to the CPI program, and whether they were a Visiting Teacher or also asked to be a Resident Teacher. This information may help the reader in deciding which teacher story is close to their own.

Table 7. Sixth grade teacher summary

Name	Years teaching	Grades in school	Prior computer usage	Overall change	CPI teacher
Paula	3 years	K – 8	Some experience	Little	VT only
Clara	6 years	5 – 8 to be K - 8	Some experience	Very much	VT only
Barry	28 years	5 – 8	Some experience	Some	BothVT & RT
Ernest	12 years	5 – 8 to be K - 8	Some experience	Very little	VT only
Shelly	3 years	5 – 8	Some experience	Very much	VT only
Molly	3 years	5 – 8	Some experience	Very much	VT only

Paula

Paula taught grade six. When she participated in CPI in 2000, she had been teaching for a total of three years, only one year at this large Kindergarten to eighth grade elementary school in northern Philadelphia. This school was the same school where Darrell, Edie, and Dee taught third grade (see Chapter VIII), Gina and Jo taught second grade (see Chapter VII), and Ellen taught Kindergarten (see Chapter VI). This school's student population had over 50 percent African American, 24 percent Latino, 18 percent Asian, and 5 percent Caucasian. Over 70 percent of the students were from low-income families and eligible for free breakfast and/or lunch. According to Paula, there are many languages spoken at this school, with the largest second language being Spanish (anonymous, personal communication, December 7, 2004).

Paula participated in CPI to get a computer and printer for her classroom. She said the principal of the school encouraged her to participate. She already knew how to use a computer for personal needs, but saw a great opportunity to get a computer and printer for the classroom.

I really enjoyed going to the classes on Saturdays. There was so much to learn about the software and the computer. I learned about the hard drive, different software and how to access materials off the Internet. I visited a teacher at a middle school as well, where the skills I learned through CPI were reinforced. I didn't have a car at that time, so the director of CPI picked me up and took me to more than one school to observe teachers. I felt I learned more because of this.

Before CPI, one of my biggest concerns about using the Internet with my students was time. I realized the computer was a very useful tool in the classroom and I wanted the students to use it often. My concern was that the students would search for a topic and spend so much time looking for information and not actually get anything out of the search. Personally, I find myself searching for sometimes hours for a single piece of information even when I use search engines, quotation marks and plus and minus signs.

My other concern was that my students would get off-task. For example, if I asked them to find out information about Roberto Clemente, they would search for unrelated topics and 'play' on the computer. When I say 'play,' I mean searching for topics unrelated to our course of study, or basically being off task. I was also concerned about 'student friendly' information. I didn't want them searching for a topic and ending up on sites that are not appropriate for children."

Through CPI she learned about kid-friendly Web portals that she could set up as default sites for her students to use when searching for information. Two specific skills she reported learning and using immediately in her classroom were how to set up the computers for students' journaling and how to cycle the students on and off the computer over a couple of days.

I had three computers in my classroom and a schedule was posted in the computer center of the room for students to know when it was their turn on a computer. The schedule was the same each week, so students quickly learned which day was their day for the computers.

As with Darrell, Edie, and Dee, Paula no longer had three computers in her classroom as of December 2004. One computer had been removed to support the new literacy programs in the school. She found having only two computers less accommodating to having students regularly use them to do meaningful work. There was also the new core curricula. She said, "Once I get this curriculum down, I'll start back to incorporating the computers. It's lock step and I'm still getting it down."

Another outcome reported by Paula was the networking that grew between the teachers. "CPI teachers would invite other CPI teachers to visit their classrooms and/or share teaching ideas. My neighbor at the school didn't know how to use a computer. So I taught her how to get on the computer. My CPI training gave me the confidence to teach another teacher how to use a computer. I would never have done that before CPI."

Paula's classroom practices changed ever so slightly to include students using the computers for journal writing and typing up projects. Perhaps this was not what some would call change, but for Paula, who had not included technology in her classroom to even a small degree prior to her experiences with the CPI program, having students use the computers at all was a change from past practices. Unfortunately, change was short lived in her classroom due to the removal of one of her three computers to support the school wide literacy initiative, complicating further classroom practice affected by the literacy and mathematics core curricula. Paula's new classroom practices were victims of the district's response to the NCLB legislation (U.S. Department of Education, 2001).

Clara

Clara had taught sixth grade math and science for six years at a middle school in southwestern Philadelphia. She taught two classes of math and science. Nearly 100 percent of the students were African American and nearly 80 percent came from low-income families and were eligible for free breakfast or lunch. This school was classified as a middle level enrollment school, down from a high enrollment in

2000. The school lost over 40 percent of the student population in six years. Low annual attendance hovered around a poor 86 percent. Student mobility ranged from a low of 15 percent in 2000 to a high of 32 percent in 2004. The neighborhood was changing, and district leadership was moving to change this middle school into a Kindergarten through eighth grade school.

Clara said she wanted to take the CPI course to get a computer for her classroom and to become more computer literate (anonymous, personal communication, May 17, 2005). She had three computers in her classroom in spring 2005. She learned about slideshows, Internet scavenger hunts, and using the Internet for research. At the time she took CPI in 2001, the school basically had computers in the computer lab, but not in the classrooms.

From CPI I was encouraged to take courses to gain computers for my classroom and I learned strategies and incentives for students to use the computer. For instance, on one older computer I have math games to help students master skills. If they complete their classroom work, they can go onto the computer and work on math skills with the games. The students typically do not have a computer at home, and time in the computer lab is limited. So I allow the students at the beginning of the day to come in and take turns using the computers for research. I monitor what they are doing, but basically they are independent. I don't have a printer in my classroom which would really help with the students. With a printer, they could print out their work.

Mostly, Clara came away from CPI with classroom management strategies for using the computers with her students. Setting up a rotation strategy for students to have a certain amount of time on the computer each week provides an incentive to get work done so they can have access to working on the computers.

When asked about individuals in the school who influenced her to take CPI, she mentioned the assistant principal (AP) from 2001 and the computer teacher.

The AP, in particular, strongly urged me to take the course. He was always encouraging the use of computers, and taking district offered courses that would bring technology into the school that we didn't have. And the computer teacher was always trying to teach us different programs and showing us useful Web sites. They both said the CPI would be useful."

As with Paula, Clara learned how to cycle 30 students on and off the computer she had in her classroom. In spring 2005 she had three computers in the front of her classroom in her computer center. Three computers were better than one for allowing student access and opportunity. Clara made time for her students to work on the computers during free times, like lunch, or before or after school. By giving them access, she sent a signal to the students that their time using the computers for academic purposes was valuable and worthwhile.

Barry

Barry started teaching at a middle school in northern Philadelphia in 1994. This was the same school where C.C. taught fifth grade (see Chapter X). He taught there for eight years before moving to another school in the district. In total, Barry had taught for 18 years by 1994. That year, 1994, was also the second year Barry described himself as being computer literate. He was using the Internet, word processing materials for his classes, graphic organizers, and Internet-based literacy programs. It is important to know that this middle school, according to Barry, considered itself a leading school with technology, so being part of the CPI pilot project was a natural.

This middle school in northern Philadelphia had grades five through eight. It was located on a busy street in a neighborhood consisting of row houses. Over 50 percent of the student body was Latino, while only 30 percent were African American and 10 percent were Asian. Over 80 percent of the students came from low income families making them eligible for free breakfast or lunch, a percentage nearly 10 percent higher than the district average. In spring 2006, the student population was classified as high, as defined for this book. Attendance was at 89 percent percent annually, and mobility was around 17 percent of the students annually.

When the principal at that time asked him if he would like to participate, Barry said "I could 'smell' the funding and support" (anonymous, personal communication, June 10, 2005). He had already seen the impact of technology in support of teaching English as a second language in general, and this was an opportunity to gain more experience and support. This middle school was a partner school with a nearby elementary and a large high school for planning and implementing CPI.

It was a real articulation opportunity in the community and the funding was definitely there.

For me, this was some of the best teaching in my career. We had the TAPS summer program out of CPI. The Technology Assistive Problem Solving program lasted for three years. The idea was for students to work on enrichment problems and collaboration with others, and for CPI teachers to see how technology could be used with kids. Two years they did projects on food safety and one year on pirates of the Caribbean. We used concepts of multiple intelligences throughout the curriculum. In the morning, the students would sit in a circle in the gym and the teachers would review the lesson for the day. They would then go to their classrooms and work on projects rotating teachers. At the end of the day they would sum up what they had learned. At the end of the summer school session, the kids displayed what they learned through a play, a dance, a song, slide shows, what ever seemed to represent their learning. They showed off to one another, to the CPI teachers, the community and to IBM.

During the school year, there were CPI teachers who visited the school and Barry's classes. He would show them how to use the computers with his students and how they could do the same back in their classrooms. A number of teachers from this school participated, which created a critical mass of teachers to support one another in using technology with kids and projects. One of the teachers in the school established a teacher discussion/study group to reflect on technology and teaching practice. As Barry noted, "collaboration is the key." Unfortunately, due to a lack of substitutes to take over classrooms, not many teachers from the middle school completed the CPI training in the end.

For Barry, the experience led him to be more involved with technology. Since he participated with CPI from the beginning, he learned from colleagues in his school. He was actively involved in the summer program that brought other teachers from around Philadelphia into his school and classrooms, with whom he could share the practices he learned through CPI and learn from them. He took this wealth of knowledge with him as he moved on to another school in the district, where once again he was a resource for other teachers.

Ernest

Ernest had taught for 12 years in a middle school in northwestern Philadelphia, and he taught sixth grade math and science. The neighborhood consisted of single family homes with trees and manicured lawns typical of a middle class neighborhood. In 2001, when Ernest participated in the CPI program, his school was a sixth-through-eighth grade middle school, but it was slated to be eliminated over time by moving the grades to a nearby elementary school. His teaching schedule varied from year to year, but mostly he taught sixth grade before fall 2005.

The total enrollment at the middle school was classified as low, as defined for this book. Over 97 percent of the students were African American, while 2 percent were Latino. Sixty-two percent of the students were classified as coming from low income families and eligible for free breakfast, which was 10 percent lower than the district average. Annual attendance was typically around 90 percent, and mobility of students in and out of the school was around 20 percent.

Ernest participated in CPI in 2001 and he was not eager to participate. As he said himself, he was perfectly happy with his teaching and classroom just as it was (anonymous, personal communication, April 28, 2005).

Basically, I participated because the computer teacher was pushing me to be computer literate. I probably wouldn't have even known the course was offered if the computer teacher had not encouraged me to participate. I describe myself as basically a non-user of computers. I didn't see the point. I was doing just fine as a teacher without computers.

Ernest indicated that he was glad the program was available and he learned a lot.

Now I have students do research on the Web, which is something I would never have considered before CPI. We go to the small computer classroom and they do projects together. Before CPI all the work was in the classroom. Now I have branched out.

Ernest was the first teacher who did not participate in the CPI program to get the computer for his classroom. Perhaps that was one of the goals for the computer teacher, assuming that if Ernest had a computer in his classroom he would make use of it with his students. Instead, Ernest took his classes to a small computer lab on the first floor of this school, when it was available and where there were 12 computers with headsets. He paired the students to perform Web research on various topics. He did not indicate that his students typed papers or journals using the computers. While this does not mean they did not, it perhaps reflects that Ernest came kicking and screaming to the CPI program, and was not exactly an example of change. But that is the point—not all CPI teachers came away from the program infusing technology into their classroom practices. Ernest chose to have the experience and continue mostly as before, since he still did not see how the use of computers would improve his classroom practices.

Shelly

Shelly had taught in Philadelphia since October of 1999, all that time at the same middle school in the southwestern part of the city. Initially, she was a substitute English teacher. She taught sixth grade reading, as well as English, to seventh and eighth grades.

Because the students at this school were not doing well academically, the school was turned over to an external management organization to run the school in 2002. As a result, the middle school student population at this school was divided by gender, so the boys were all together and the girls were all together in separate parts of the building. This school was a large middle school with enrollment classified as high, as defined for this book. Eighty-seven percent of the students were African American, while nearly 8 percent were Asian and 4 percent were Caucasian. Eighty-four percent of the students were classified as coming from low income families and eligible for free breakfast and lunch. Attendance averaged at 87 percent annually from 1999 through 2005 and student mobility in and out of the school hovered around 20 percent annually.

The school facility was a huge building with many sections. Shelly's classroom even had a balcony, as did many classrooms located on the outside of the building. Some of the classrooms looked onto the atrium with no outside view. Inside the building there was a feeling of airiness.

In 2001 Shelly's classroom did not have a computer, as was the case for the majority of classrooms at this middle school. If a classroom had one or two computers, they were typically much older models with few software options to use with the students. Shelly described her teaching at that time as team teaching, and that she looked for any teaching tools and methods to help her children learn (anonymous, personal communication, June 21, 2006). In spring 2006 she taught on the boys' side of the school. She noted that she really liked teaching the boys and found the separation from the girls to give the boys a chance to excel academically rather than "show off for the girls."

In the team teaching, they did projects. She often would go onto the computer herself to find materials and information for her children, printed out the materials at home and brought them into school to duplicate. "It was a few extra steps, but I had to do them back then." She wanted to get the children interested in the projects.

The CPI program came along and three teachers from [here] signed up. To be honest, we needed new computer technology for our classrooms. And I was a 'PC person' at that time and knew that I would need to learn how to use an Apple computer. I saw CPI as a means to both ends. I learned how to use the Apple [computer] and how to use such resources as Inspiration for graphing and concept mapping, and PowerPoint for presentation projects.

The best part was getting to go to another school to see another teacher use technology. I went to [another middle school] to another English teacher's classroom. She had six computers in that room. She had them working on projects in pairs at the computer stations. She made it look very easy, she was so organized.

After her CPI visitation, Shelly's class was to work on a project for four weeks. The project topic was the Underground Railroad, about which she was very excited. They read about the Underground Railroad and Harriet Tubman's story.

The students were to imagine themselves as slaves and they were on the Underground Railroad to safety. They were to write a narrative about themselves to set the stage. They did cross-curricular activities that brought in science in talking about the weather and math in calculating how long it would take them to walk from point A to point B each day, plotting their progress on maps, bringing in social studies. Each day I would give them a journal assignment with a specific set of circumstances for them to hypothesize or write about how they would deal with these issues. It was really cool!

In 2006 Shelly had three computers in her classroom: one Apple, one Dell, and a new Apple laptop, but no printer. Without the printer, doing work on the computers was cumbersome and vaguely pointless for the students. So, if she had work for them to do on a computer, she took them down the hall to the computer lab and they worked there where it was more spacious and supplies and equipment were available, especially printers.

When asked how CPI impacted her teaching, Shelly said:

It may not have impacted my teaching as much as other teachers. I was always looking for ways to get resources for my students. CPI gave me another option. And the teacher I visited gave me resources and Web sites I now use to supplement the curriculum. The curriculum also is an impediment, to some extent. The lessons are pre-planned and must be followed on the pre-set schedule. This really makes it hard for me to be creative in using technology. Including such projects as the Underground Railroad is much more difficult unless it is already in the literacy curriculum, and team teaching across subject areas has become difficult as well.

Shelly's classroom was a spacious room. When you came in the room, to the right were large chalkboards, as well as on the opposite wall. In between was a bookcase, and there was a ledge in the back of the room where she kept textbooks when they were not in use. Tables were set up in groups of four so students could do collaborative and cooperative activities, and there were tables in the back of the room with her computers and space for working in small groups. Shelly did not think her room was spacious but that she used the perimeters of the room to the greatest advantage, making the room seem spacious. All in all, it worked for her and her students.

In the end, Shelly's classroom practices did not change very much. She continued to include computer work in her teaching, but the lack of a printer in her classroom and the core curricula gave her reasons not to include much. Shelly already was predisposed to search out learning tools for her students before she experienced CPI. CPI added to her tool-kit of options for engaging her students. However, circumstances beyond her control made it difficult for her to bring those options to bear on the curriculum and her students.

Molly

Molly participated in the CPI program with Amanda, another teacher from her school who taught seventh grade (see Chapter XII). She had just started teaching in the district in 2002. While she was prepared as an elementary education teacher, she was anxious about teaching in a middle school computer classroom. The 15 hours of training occurred in a school lab, and the trainers were very helpful in giving examples. Molly noted she was able to observe how another school lab could be managed with 17 "needy students" (anonymous, personal communication, June 12, 2006). While Molly taught fifth, sixth, seventh, and eighth grade students, we talked mostly about the sixth grade.

Molly taught at an academic magnet school where students applied to be admitted, so there were students from all over the city. Enrollment in spring 2006 was classified as high, as defined for this book, and comprised of 47 percent percent Caucasian, 30 percent African American, 15 percent Asian, and 8 percent Latino students. Annual student attendance was 96 percent, student mobility was less than 1 percent, and only 40 percent of the students were classified as from low income families. This was clearly not the typical neighborhood school.

Observing the eighth grade English teacher at another middle school was also really helpful for Molly. That teacher had three computers in her classroom, all because she had been involved with the CPI program. She had a schedule for students to use the computer and a timer on the computer. The students knew exactly what to do and what the schedule was. It was very organized. This teacher also introduced Molly to the BrainPOP® Web site. "This site contains seemingly endless information for students and teachers. It includes experiments that cross all subject areas and short, cute movies."

From my college experiences, I was already prepared to do the various projects assigned through CPI, but being able to be successful gave me a boost of confidence. Especially learning how to manage a computer lab was most helpful for me.

As for how CPI impacted her teaching, Molly said it had a significant impact on her teaching. It was a confidence booster, for sure. She gained many examples of ways she could teach in her own classroom. Examples included taking a floppy disk apart to see what it looks like inside and taking an old computer apart to see what the parts look like. These were things middle school children love to do!

Since CPI, Molly had become a resource in her school for other teachers. She had given lessons on developing a database of resources to reflect the curriculum that students could use as part of a lesson. She described one teacher who used a "fishbowl" teaching method and the computer in the classroom was one activity contributing to the lesson.

Molly noted that you can teach a teacher how to use a particular software application and describe how it might be used in their classroom activities, but—until they really have a need for it—they would not actually use the software. And if they do not use it regularly, they will forget how to use it. Molly gave workshops on how to use Inspiration and PowerPoint. She loved it when her students learned a software application, like PowerPoint, to create a learning game and then used PowerPoint to create another game for another course.

She was also very excited about the new instructional technology curriculum mandated for fall 2006, like Parker noted in Chapter VI. When she started teaching, no computer curriculum existed and she went on her own to the International Society for Technology in Education (ISTE®) Web site and used the student standards from there. The new technology curriculum was based on these standards, so Molly was a bit ahead of the game. However, the fifth graders in her school did not have computer class. This meant she would have sixth graders who were not prepared for the curriculum because it presumed prior knowledge. And, she only saw the students for a limited period of time, once a week. Achieving all the standards would be a challenge, but Molly was up for it.

Molly was the first teacher presented in this book who was a computer teacher when she participated in CPI; however, she was also the teacher with the least teaching experience. Many of the teachers presented in previous chapters had taught 20 or 30 years and were looking for ways to engage their students that might also give them ideas for new ways to teach what they had been teaching for so long. For Molly the CPI program was a way to learn how other teachers taught with technology and to give her confidence that she could be a computer teacher. She was very comfortable with her training to be an elementary school teacher, but not so much with the computer classroom. CPI gave Molly a different set of skills than other teachers, and she became a resource for her school, much as Barry did.

Summary

Here were stories from six different sixth grade teachers. All but one was from a middle school environment. All but two were regular classroom teachers. All but two participated in CPI initially to get the computer for their classroom. In the end, all added experiences to their classroom practices that were not present before participating in CPI. However, having said that, the new experiences were minimal by being limited to searching for information on the Internet or typing stories. Barry, perhaps, was the one teacher who truly saw how computers could support the ESL programs. Clara made access to the few computers in her classroom open to allow her students opportunities to explore the world of available knowledge. Molly was going to be teaching with computers anyway by virtue of being the computer teacher, but she had no curriculum and was looking for support and resources to help her have the confidence to do well.

Two teachers, Paula and Shelly, described circumstances that precluded and prevented their use of computers within their classrooms. If they needed an excuse to not use the computers, their schools gave them ones. By taking the computers out of the classroom, as was the case for Paula, the school made it more difficult for her to use the computers remaining in her classroom with her students. Not being able to print student work in the classroom made Shelly retreat to the computer lab, assuming it was available when she needed it. In either case, the lack of use by these two teachers was not as expressed by Ernest, who never seemed to quite accept the notion that computers could be helpful in his classroom.

Analysis

Change is difficult at best. As noted by Fullan (2001), change in classrooms and schools takes support from many levels. These sixth grade teachers did change, but not as much as teachers presented previously in this book. Unlike teachers from Chapters V through X, these teachers did not speak of collaboration with other teachers in their schools, except for Barry when talking about the TAPS CPI summer program in the late 1990s. They talked about how wonderful and educational it was to visit another teacher's classroom, but they did not talk about supportive activities within their own schools. It was interesting that Ernest appeared to have participated only because he was intimidated by the computer teacher, and then he made only perfunctory changes in his classroom practices. Fifteen teachers from that school participated in CPI and two of them were Resident Teachers who regularly had CPI teachers visiting. From previous chapters one would assume the environment was ripe for collaboration and support across the school; however, this assumes all teachers truly accepted computers as necessary for their classrooms.

Zhao and Cziko (2001) noted that teachers who do not embrace technology may have simply decided that they saw no reason to change their practice. This approach to non-change could be applied to Ernest. He said he was perfectly happy with his classroom practices before participating in the CPI program. If you looked at his classroom practices after CPI, there was minimal change from before the experience. Clearly he participated to get the computer teacher off his back, rather than from a reasoned decision that computers could provide a set of resources to his classroom that he did not have. This is not to condemn Ernest for not embracing technology more fully, but to hypothesize on why he chose to minimize use of the skills and knowledge he gained or to which he was exposed through the CPI program.

Using Rogers' diffusion of innovations theory (1962, 1972, 1983, 1995, and 2003) as a lens, each of these six teachers started the CPI experience at different stages of adoption. All had awareness of computers, and none of these teachers classified themselves as novice users. Ernest was a non-user, but he knew how to use basic functions of a computer. Paula, Barry, Clara, and Shelly saw CPI as an opportunity to learn more about computers and how they might impact their classrooms and, therefore, decided that the CPI program would help them determine how computers would fit into their classroom practices. All of these teachers tested the option of using computers in their classrooms through observing the Resident Teacher to whom they were assigned; and all but Ernest decided to implement computers into their classrooms at some level.

Barry and Molly took their experiences through CPI to another level entirely. They became computer resource people for their schools, working with other teachers to help them consider how computers could provide solutions to various classroom problems. They took it upon themselves to be change agents in their schools, rather than to merely benefit from the CPI program alone.

The Concerns Based Adoption Model (CBAM), by Hord, Rutherford, Huling-Austin, and Hall (1987) gives another dimension for analyzing these six teachers. All of these teachers, at one point or another, reached the fourth stage of adoption, or "consequence." They all were concerned with how use of computers could impact student learning. Ernest appeared to have decided that student impact was inconsequential and thus chose not to use computers for anything regularly. Paula also reached stage four until the school removed one of her computers, which moved her back to the third stage, or managing the time for students to access the computer meaningfully. Shelly moved backward to stage three because of the lack of a printer. Only Barry and Molly moved into stage five, or "collaboration," by working with other teachers in their schools on technology solutions.

CBAM also provides a look at the levels of technology use as a lens for analysis. Only one of these teachers described himself as non-user of technology. They were well aware of their personal need to learn more about computers, with Shelly noting that she was a PC user and needed to learn how to use an Apple computer. Their decisions were personal, not related to the curriculum. Even for Molly, who was going to be the computer teacher at her school, her decision to participate in CPI was to help her personally be more comfortable teaching in a computer lab. Only Barry and Molly were routine users of computers in their classroom. Paula was initially, but by 2004 she was focusing on mechanical uses of the computer rather than how to infuse it into her daily practices.

In Fullan's 2001 book *The new meaning of educational change*, he notes three interactive factors that affect implementation of an educational change in a school or classroom: the characteristics of the change, the local characteristics, and external factors (see Chapter XI). For these six teachers, the characteristics of the change, that of bringing a computer into their classroom practices, was clear, but the need for the change was not consistent across individuals. All but Ernest apparently understood the need for change in practice to include computers as part of classroom practices. Ernest was never convinced that this need was real, rather than something manufactured by the computer teacher in his school. The local characteristics of the school infrastructure and programming impacted Paula and Shelly by making it difficult for them to implement computers effectively in their classrooms—or at least that was their perception. External factors appear to be negligible, unless they were influenced by the political nature of technology in classrooms. None of these teachers spoke to any external pressure on them to learn about technology, other than Ernest. Even Ernest did not speak of any influence other than the computer teacher, so he was not influenced by any national pressure that computers be infused into classrooms, as was expressed by teachers in Chapters VII and IX. For Ernest, there simply was no pressure for him to embrace computers into his classroom, except to accommodate the computer teacher in his school.

Lastly, the ACOT levels of adoption of technology into classrooms were helpful in determining how much these teachers had truly integrated computers into their practices. (Fisher, Dwyer, & Yocam, 1996; Gordon, 1996) Only Molly and Barry started their CPI experiences well beyond the entry level and perhaps already moving toward the innovation level. The other four teachers were relatively anxious about how they would use technology with their students, including word processing and research. Did these four teachers move forward? Paula, Clara, and Shelly did move into stage two, or the adoption level. They were using computers for productivity for themselves. Each was moving toward level three, or adaptation, but not reaching that level fully. Clara encouraged her students to use the computers in her classroom when they had the time, but she did not regularly make time in the curriculum for student use.

In the end, what was the level of change for these teachers? For Ernest the answer is none since basically his overall classroom practices did not change. He would perhaps argue that, because he occasionally had his students do "research" on the computer, he changed his practices; but that "change" was not great, happened only rarely, and was far from infusing computers into classroom practice. For Barry and Molly, change was not so much in their own classroom practices as becoming a resource for other teachers in their schools. Paula, Clara, and Shelly changed their practices, but not as much as some critics of technology use in classrooms would suggest. Before their experiences with CPI, only Barry and Molly were regular users of computers. After CPI, each of these teachers saw the value of computing in empowering their students to both take control of their learning and in the expression of knowledge.

References

Fisher, C., Dwyer, D. C., & Yocam, K. (Eds.) (1996). *Educational and technology: Reflections on computing in classrooms.* San Francisco: Jossey-Bass Publishers and Apple Press.

Fullan, M. (2001). *The new meaning of educational change.* Third Edition. New York: Teachers College Press.

Gordon, A. R. (1996). *Computers in the classroom: How teachers and students are using technology to transform learning*. San Francisco: Jossey-Bass Publishers and Apple Press.

Hord, S. M., Rutherford, W.L., Huling-Austin, L., & Hall, G. E. (1987). *Taking charge of change.* Alexandria, VA: ASCD.

Rogers, E. M. (1962, 1972, 1983, 1995, & 2003). *Diffusion of innovations*. Glencoe, IL: The Free Press.

Scott, T., Cole, M., & Engel, M. (1992). Computers and education: A cultural constructivist perspective. *Review of Research in Education, 18,* 191-251.

U.S. Department of Education. (2001). No Child Left Behind Act. Washington, D.C.

Zhao, Y., & Cziko, G. A. (2001). Teacher adoption of technology: A perceptual control theory perspective. *Journal of Technology and Teacher Education, 9*(1), 5-30.

Chapter XII

Becoming Tech Experts:
Grade 7

Computers are not an end in themselves. The goal of technology integration into the classroom and curriculum is not to expose students to computers and the Internet. Technology, by definition, is a tool. That tool should be serving a greater purpose, and what that purpose is for you may be very different from what it is for the school down the street (Dockterman, 1998, p. 21).

As with fifth and sixth grades, the seventh grade classroom depends on whether the school is an elementary school or middle school. In many Kindergarten through eighth grade schools in Philadelphia, seventh graders have two different teachers rather than only one as with sixth graders. One teacher concentrates on literacy and social studies, while the other teacher takes on mathematics and science. These students cycle between two different classrooms. In contrast, students in middle schools may have a homeroom teacher, but they cycle through a number of different classrooms and teachers for each subject. Their school experiences are much different from those of students who only travel between two classrooms.

In addition, seventh graders are typically 13 or 14 years of age. They are growing by leaps and bounds physically and emotionally, which challenges the teachers who are trying to help them grow intellectually. Keeping these students attention attuned to what is happening in the classroom, rather than what is happening in the hallway, lunchroom, etc., becomes a never-ending process. Computers in the classrooms can help.

Piaget's stage of development for this age range is the "formal operations" stage (1972, 1990). Here children begin to move into abstract or adult thinking. Not all children have developed to this stage and may continue to develop in the "concrete operations" stage where they cannot yet think in abstract terms. The seventh grade classroom brings a mixture of students' intellectual development, again challenging the teacher to present materials in both concrete and abstract terms to adequately engage and promote learning by all students. Computer simulations and productivity programs can support efforts in both directions.

This chapter presents six seventh grade teachers from different parts of the school district. All but one of these teachers taught in a middle school setting. One teacher went on to become a Resident Teacher. Their experiences with CPI were different, and their paths to using technology in the classrooms were divergent. Each has a unique story that represents many of the approaches to using technology in middle grades.

Table 8 gives the reader an overview of the teachers presented in this chapter. The information provided includes how many years of teaching experience they had prior to participating in the CPI program, what grades were included in their school, how they characterized their prior use of computers/technology prior to the CPI program, and whether they were a Visiting Teacher or also asked to be a Resident Teacher. This information may help the reader in deciding which teacher story is close to theirs.

Table 8. Seventh grade teacher summary

Name	Years teaching	Grades in school	Prior computer usage	Overall change	CPI teacher
Sandra	19 years	5 – 8	Novice user	Very much	VT only
Amy	5 years	5 – 8	Novice user	Very much	VT only
Amanda	10+ years	5 – 8	Novice user	Little	VT only
Elaine	20+ years	K – 8	Some experience	Some	Both VT & RT
Sam	20+ years	5 – 8	Some experience	Very much	VT only
Nancy	26 years	5 – 8	Some experience	Very much	VT only

Sandra

Sandra taught math and science at a middle school in southwestern Philadelphia, the same school where Clara taught sixth grade (See Chapter XI). For a few years she taught eighth grade students, but she was teaching seventh grade math and science when interviewed. She had been at this middle school for 13 years, and had taught for 19 years. She participated in CPI in 2001.

When asked to describe her school, she said she loved it with all the programs and opportunities for students and teachers. She taught a computer refurbishing class for students. Basically, this was a middle school with grades six, seven and eight. The school was on a lovely street in southwestern Philadelphia with single family homes and row homes. There were businesses interspersed in the neighborhood, and the trolley ran right in front of the school. According to Sandra, this was one of the middle schools slated to become a Kindergarten through eighth grade school over time (anonymous, personal communication, May 17, 2005).

In spring 2006 the enrollment was at the middle level as defined for this book, which was down 46 percent from a high enrollment in 2000. Students came from a variety of backgrounds, but nearly 98 percent were African American. Eighty percent of the students were from low income families. Annual student attendance between 1999 and 2005 averaged 88 percent, below the city target figure and student mobility averaged 18 percent.

Sandra's classroom was a corner room on the second floor. Her five computers were along the wall upon entering the room. The room was full of individual student desks that could be set in a row or grouped for collaborative activities. Her spacious classroom was full of science projects and materials to support the science curriculum.

When asked why she wanted to take the CPI program, Sandra said she wanted to take it to know about computers (anonymous, personal communication, May 17, 2005).

Basically, before CPI, I would classify myself as a non-user. I knew nothing about computers or using them in teaching. Now I have my students using the Math24 games, using the Internet for research for math and science projects, writing papers, and using PowerPoint for presentations.

I would like to work toward a paper-free environment. I currently have five computers in my classroom and a printer. I went from being a non-user of computers to co-teaching the computer classes with the computer teacher, teaching a computer refurbishing course, and teaching students about robotics. I know I have been an inspiration to my students. There is this one student who does not have a computer at home, but has become the local school computer whiz fixing computers and troubleshooting. He's really gained in self-esteem.

When asked who inspired her to participate in CPI, Sandra specifically mentioned the 2001 assistant principal. He was an inspiration for wanting to learn about technology. "He saw the potential and the future in technology in classrooms."

Since CPI, Sandra has been a resource for her colleagues as well as her students. When the teachers have questions, they ask her for help. When the students have questions, they count on Sandra to point them in the right direction for them to be self sufficient. This is a truly wonderful story of going from being a non-user to being a resource for others in the building, teaching how to refurbish computers and robotics to students. For some CPI teachers, the challenge was to learn more and do more. Sandra is one example of someone who took up that challenge to go further in their own knowledge and expertise.

As a science teacher, Sandra embraced the idea that technology can support students learning about science. Robotics supports the physics component of seventh grade science, and achieving with robotics gives students a level of fulfillment that is hard to gain elsewhere. One year the students were invited to the central computing offices where all the programs teaching robotics showed what their students could do. It was not a competition, per se, and the students learned from each other. That session was really empowering for students, and Sandra made that happen by being open to how technology could support her teaching and how robotics could improve her students' sense of self.

Amy

Amy was a seventh grade social studies and language arts teacher. She had taught for more than five years at the same school where Clara taught sixth grade (see Chapter XI) and Sandra taught seventh grade in southwestern Philadelphia. Amy participated in CPI in 2001-2002 because she wanted the opportunity to integrate technology into teaching. Amy described herself, like many CPI Visiting Teachers, as a non-user before CPI (anonymous, personal communication, May 17, 2005).

I wanted to be able to assign more computer work for my students and to be more comfortable myself with computers. At that time, I had some tutorial games and CDs for drill and practice, but mostly computers were in the computer labs.

She learned not to be rigid using only the book. Lots of things could be done on the computer to help students learn more.

Students now stay in my classroom, or come back early from lunch or recess, to work on the computer. They are making brochures, using PowerPoint, and doing research for projects on the Internet. I do watch where they go, but they are being productive and learning.

The day Amy's classroom was visited the three computers were in use for a social studies project. Students were actively engaged in searching the Internet for information and images. They were to put the information together, and to synthesize it into a presentation using PowerPoint to demonstrate their learning. Amy noted that she would not have had her students do that type of presentation before CPI, because she simply would not have had the skills or the courage to try. Now she had them doing PowerPoint presentations regularly.

The success story here was a teacher who was a non-user before starting the CPI program who now had a computer at home and three computers in her classroom. She no longer classified herself as a non-user and regularly engaged and helped her students use computers to support their learning. She noted the computers were used every day, and she learned how to rotate the students on and off so everyone had time to accomplish their goals. With three computers and over 30 students, this was a challenge, but she learned classroom management strategies to help her accomplish her goals with her students through CPI. She learned from her Resident Teacher to establish a schedule for the week. That schedule did not have to be the same for each week, but could change as curriculum priorities changed. Students knew their assigned time each week and made sure they got their work done on time, or they could lose their computer privileges the next week.

Amanda

Amanda taught world geography to seventh graders and African American history to high school students at one of the district's magnet schools, the same school where Molly taught (see Chapter XI). She had taught there for a total of five years. Prior to this school and this district, Amanda taught middle grade social studies for seven years in one of the regional suburban districts.

In describing her teaching style, Amanda said she tended to be a hands-on, interactive teacher, "With middle grades, I believe they need to be engaged in the material, so I try to go beyond lecture and notes" (anonymous, personal communication, June 1, 2006). Having said that, Amanda did not have a single, regular classroom where she taught. One way to describe her access to classrooms was to say she was a "migrant teacher" within her own school. She had access to technology on a mobile cart. But, as she said, "I take my materials between classes on a cart as it is.

If I want to use a computer and projector, I have to go get another cart." This was quite cumbersome to say the very least.

Amanda wanted to participate in the CPI program because she wanted to get a computer for her classroom when she finally got a regular classroom. In addition, she considered her technology skills as "so-so;" therefore, the technology training itself was of value to update her computing skills. She hoped to learn to use technology in her classroom teaching, even if access was a bit difficult. Amanda noted that at the CPI Saturday training there were a number of other teachers who were technologically challenged. By this she meant there were teachers who needed to learn how to turn the computer on and how to use the mouse. These teachers took up a great deal of time for the instructors, leaving those who had some technological skill level to explore software on their own. Instruction was given, but the neediest teachers got more attention. This should have been taken care of by the self-assessment survey, but clearly some teachers got into sections where they should not have been placed.

I was interested in learning how to use Inspiration and how to create my own Web site, neither of which happened for me through CPI. I learn better through hands-on practice with someone to demonstrate and show me steps. This didn't happen for me.

The computer she received as a result of the CPI program was still with her at the school, but she could not use it in her teaching, per se, other than to create PowerPoint presentations and to use it to research the Internet for materials to support her curriculum. As noted above, she did not have a dedicated classroom for her teaching, but moved from room to room, floor to floor, as her students did, so her CPI computer stayed in a central location. Amanda noted she could spend hours finding materials on the Web to bring into presentations. The school became a wireless environment in spring 2006, so when she did have a computer and projector to use, it would be easy to incorporate Web site links into the presentation and bring the materials alive through the Internet.

My Resident Teacher was at a middle school in northwest Philadelphia. There were two Resident Teachers at that school, but my RT was using technology mainly for writing. Unfortunately, when I was visiting she was not able to provide guidance on incorporating technology into the classroom. She told me I could stay in her classroom if I wanted to, but she was not using technology that day. So I spent time going through software and Web sites. Although I saw some good stuff, a lot of it made me feel really inadequate—like there is so much out there and I was so far behind. I think the essential thing is to think in terms of small steps, adding and incorporating skills one at a time and looking only at myself as a barometer of my own progress. It was helpful to have an RT who could understand some of my feelings about that.

This experience was not unique to Amanda, since it is true that technology is not used all day, every day in any classroom. Unfortunately, the two Resident Teachers had not coordinated a series of experiences for Amanda so she could gain more knowledge than she did.

Not having a classroom of her own hampered Amanda's ability to explore and expand on what she learned through the CPI program. Her school gained a mobile laptop lab in 2006, and each middle grade teacher received a new wireless-laptop to access the network and the district's instructional management system. In fall 2006 this school was scheduled to receive a few "smart classrooms" with smart whiteboards available for teachers to use. Amanda noted that her department chair would have one of the rooms as his regular classroom, but it could be used by other teachers on an as-needed basis.

Amanda benefited from CPI because she got a computer for her use in the school. Prior to CPI she had access to a computer, but it was not one dedicated to her use. Remember, she was a "migrant teacher," moving from classroom to classroom all day long. Given the need to move teaching materials with her, moving technology was not a priority. She said she still wanted to create a Web site for her classes and to further learn how Inspiration could support her teaching.

Elaine

Elaine taught seventh grade math and social studies at the same school where Geri, Kim, Margaret, and Becca taught third grade (see Chapter VIII for demographics of this school). In 2005, Elaine was the computer and seventh-grade mentally gifted teacher. One of her responsibilities was to schedule the distribution and management of the mobile computer labs throughout the school.

Elaine said she learned about the CPI program from a posting in the office (anonymous, personal communication, April 6, 2005). Teachers were asked to write a summary about why they wanted to participate, and the principal picked from those who applied. She thought five were picked because only five applied. When asked why she wanted to participate in CPI, Elaine said:

First, to be honest, was to get the computer that really worked. I was a computer major in college for a while, so I do have a high interest in technology and hav-

ing seventh graders, trying to get them and get myself organized in a way that we could use the computer in the classroom. Without [CPI], I wouldn't have known where to begin.

Also, technology seems to be on the rise here, because we seem to be getting more computers. They had been talking about the tech lab and keeping our computer room. The principal's working on sending us anywhere we can get computers. It's here, it's just, can we all use it when we all want to in the right way?

Since Elaine was a computer major for while in college, the CPI Saturday program was not helpful for her, but she noted how helpful it was for others who were afraid of using the computer. What really was relevant for Elaine was the Resident Teacher.

Her homeroom, when those kids came into the classroom, they would sit down at the computer. They created the school homepage and Web site. It was before the Read180®, and she used a literacy program with her low readers. The kids would come in and they would do what they needed to do. She had five computers in her classroom. She allowed me to use one of the computers, but the kids had a schedule and they knew what they had to do.

After CPI, Elaine stopped making copies of assignments and papers for her students. She said, "Teaching seventh grade, especially when I did the pharaohs and pyramid projects, I would give them one set of directions and tell them, 'You get one and that's it. If you lose it, you have to go on the computer to get the directions.' And then I would make slideshows with the directions and the whole activity. It's still up there [on the CPI Web site] with all the information. If they wanted to research while they were there, that was okay too."

Elaine was already well versed and comfortable in using computers for her own needs, but she learned how to manage her classroom to help those students who wanted help.

For the few that have been on the computer—well to be honest, the good ones went on the computer. I mean I can see them working independently and not being bothered by other students. They seem to enjoy that because my class has just been—my class is rough. These kids couldn't care less about getting on the computer. It wouldn't increase their attendance or increase their knowledge because they are seventh grade and they think they know everything. For those that were on the computers, just the fact they were sitting on there themselves or with a friend, it seemed to work well. They like the fact that they have a computer in their classroom.

Because Elaine was comfortable using computers, she became a resource for the other 11 teachers from her school who participated in CPI. Before she became a Resident Teacher, she noted that she was the de facto Resident Teacher in the school, since all the other CPI Visiting Teachers deferred to her. As noted above, in 2005 Elaine was the computer teacher and mentally gifted teacher for the school. Clearly, her role in the school changed over time.

Did Elaine's classroom teaching practices change? She stopped copying directions for projects for students who lost their directions. She started creating presentations to support project directions. She had her students create presentations to demonstrate their learning. So the short answer is yes, her classroom practices changed slightly. Her knowledge about computers and what they brought to the classroom did not. She already used the mobile computer lab available in the school. In fact in 2005, when she was interviewed, she managed the three mobile labs for the entire school, helping other teachers use the systems by turning their regular classrooms into temporary computer labs. She continued to be a technology resource for her colleagues who seemed to defer to her expertise.

Sam

Sam taught in the Philadelphia school district for at least 20 years, of which 10 of those were at a middle school in the eastern part of the city and the other 10 years at other schools in the district. He taught in a special education resource room in 2006, but taught science and reading in 2001 when he participated in the CPI program. He taught mostly seventh grade.

The school where Sam taught was a large middle school with an enrollment classified a high for the purposes of this book. The student population was diverse, with 44 percent Caucasian students, 36 percent African American students, 14 percent Latino students, and 4 percent Asian students. Only 63 percent of the student body was classified as from low income families, well below the district average. Student attendance averaged over six years to be around 88 percent, which was below district requirements. Student mobility in and out of the school over six years averaged 17 percent.

Sam indicated that he was very familiar with computers, starting out with Atari computers back in the late 1970s (anonymous, personal communication, June 19, 2006). He said he was more familiar with PC-type computers but now had Mac-type computers and wanted to be more familiar with Macs. He was very interested in how computers could support writing.

He and another teacher from the school visited a writing lab at another middle school for their rotation.

There we saw students engaged in the writing process through computers. They would discuss a topic or writing genre and do a pre-writing exercise to get ideas flowing. Next the children used a specific writing program to draft their papers, edit them, and produce a final product. While this all seemed natural, using the software made the process much easier all around than trying to do the drafts, edits, and more on paper and pencil only. The teacher and lab were well set up with networked computers. The students did a lot of work, but were very productive in the end.

While Sam was already using computers in his teaching, he borrowed ideas from the writing lab experiences. He used computers more and more with the children, especially his special education resource room children. He had them write papers first with paper and pencil and then type them. For those children who had a problem typing due to disabilities, he had one of the other children type the paper for them so they were not left out of the process. He had two eMac® computers, one PC-type computer and an iBook laptop. He also had two other computers, but they were really slow and only used for math and reading programs. He recently had a couple of Atari computers he used, but there came a time when he personally could not keep them functioning.

One aspect of CPI that Sam said was important was the collaboration between the teachers who took the program. Only two teachers from his school took CPI, but they met teachers from other schools in the district. The Saturday training they had was at an elementary in northeastern Philadelphia. They met other teachers from the area with similar interests and experiences and shared with each other. From CPI they gained the computer, printer, and grade-specific software.

Overall, according to Sam, CPI was a great program. He learned some new software programs, like PowerPoint. They talked about the one computer classroom and how to make the most of that one computer. They talked about not having just one student on a computer, but having groups or pairings of students and having them do collaborative work. Collaborative work was a much more powerful concept than each student doing research individually and not sharing out to the whole class. He also learned about Inspiration and use of spreadsheets. All of these options were extremely useful to Sam.

Again, did Sam's classroom practices change? Yes, they did. He started using the computers to support his students' writing. Sam was really interested in how computers could help his students write better, which was why he requested to visit a writing lab. He came away with ideas for using the computing power he had in his classroom to support the pre-writing activities, early writing, and final editing of work. So CPI satisfied Sam's goal to support writing.

Nancy

Nancy had taught for 26 years at the same middle school in the school district. She taught special education across all the grades in the school, but mostly talked about seventh grade. She taught at the same school where Shelly taught sixth grade (See Chapter XI for demographic information).

Before CPI, technology use by Nancy was limited. She may have had one older computer in her classroom, an older Macintosh not Internet accessible. Nancy said that technology really did not have much impact in her classroom. (anonymous, personal communication, June 7, 2006)The children could use older drill and practice software for mathematics and reading. After a lesson they could use the computer to support the lesson. They did journal writing and paragraph formatting, and they printed out materials they typed.

CPI was an opportunity to learn more about computers. She remarked, "I didn't want to always have to go find someone to help me. I wanted to learn the terminology. I wanted to be more self-sufficient with technology." Even learning the computer terminology was liberating for Nancy, because she could talk with the computer teacher and get across what her concerns were. Language can be very empowering.

It was wonderful! We learned about PowerPoint. We learned to create a Web site for children to go on and get lessons. And there was the Learning Village site where you could familiarize with other assignments and get materials or resources relevant to your classroom. It was great that way.

CPI also afforded Nancy the opportunity to become more comfortable with resources on the Internet. She was much more comfortable accessing materials and finding resources from the CPI program. The Internet could be both a strange and wonderful place at the same time. Making sure the middle school students were "safe" on the Internet was a major goal for Nancy. So learning about kid friendly Web portals helped her make sure her students could find the information they needed for reports, papers, and projects from sites where they would not be introduced to inappropriate materials.

Nancy said she found the BrainPOP Web site and recommended it to the principal as a resource for the entire school. She said, "There are resources there for all grades and subjects. Kids love it. The materials are for low level or the challenged student. It's friendly for them and easier for them to use and they do. I even gave the site for some of the parents for the summer to get their kids on and work on skills." CPI opened her horizons to many new ideas and gave her confidence for getting on the Internet. It made her more apt to check things out which she would not have done before.

Before CPI, Nancy used computers for basic drill and practice activities that supported the curriculum. After CPI she was willing to branch out and explore options on the Internet that could support the school across the curriculum. Not that Nancy was shy to do that before CPI, it is just that she was unaware of the vast resources available through the Internet. CPI opened her eyes to possibilities that supported her students and others within her school. Since she had spent her entire teaching career at the same school, Nancy had a devotion to the school and the students that comes with longevity. Helping those children succeed was important to her, and CPI helped her learn how to better achieve that goal.

Summary

These six teachers were from vastly different areas of the district. Sandra and Amy were from the same school, and Nancy's school was also in the southwest region of the district, but these were very different schools. Elaine was the only teacher from a Kindergarten through eighth grade school and the only teacher who mentioned having access to mobile computer labs. Amanda taught at a magnet school, very different from the other schools, which was just getting a mobile computer lab.

CPI touched these seventh grade teachers in different ways. Amanda was not well served by the program because she wanted to learn how to create a Web page and that did not work well for her. Elaine was already very well-versed in computing, having been a computer major in college for a while, but she learned about allowing students to take responsibility through computing. Sam was mostly interested in how computers could support the writing process and came away from CPI with ideas he implemented in his classroom. And Sandra caught the computer bug such that by spring 2005 she was teaching a class in refurbishing computers and another on robotics.

Three of these teachers initially characterized themselves as non-users of computers before CPI. Sam and Elaine had lots of experience, but were not implementing the use of computers in their classrooms. Nancy wanted to become more self-sufficient so she would not be dependent on the computer teacher every time something went wrong. All of these teachers, with the exception of Amanda, achieved their goals for participating in CPI—learning more about computers and how to use them in their classrooms, becoming more self-sufficient and confident about their own skills. Amanda was the one teacher who described her experience with her Resident Teacher as unsatisfactory because the Resident Teacher was not using computers the days she visited. While it is true that computers are not used all day long by students, she missed out on observing students successfully integrating computing into daily activities, or teachers infusing computing into teaching practices. All of which meant that Amanda came away from her visitation experience without images for technology use in classrooms.

Analysis

These six teachers came from very different environments, including the two teachers from the same school, and they came into the CPI program with different goals and needs. Elaine really did not need the computer training component of the program and would have been satisfied with visiting a Resident Teacher or two to get ideas for integrating computing into her classroom. Sam also was proficient with using computers but wanted to know more about how computers could support the writing process in his classroom. Sandra and Amy both considered themselves non-users of computers before they participated in CPI and graduated to wanting to do more to support their students. Sandra went on to have five computers in her classroom and co-teach with the computer teacher. Amy supported her students by opening up her classroom for them to come in during their free time to work on projects while the computers were not used for other purposes. Nancy wanted to learn more about computers and went on to seek out Internet resources that could support the students in her entire school, not just her classroom. Only Amanda came away from her CPI experience disappointed. She also came from a school where only one other teacher participated in CPI and that other teacher was the computer teacher. And, Amanda did not have her own classroom but moved from classroom to classroom throughout the day, making it quite difficult for her to incorporate any technology into her teaching practices.

One aspect about Amanda's school should be pointed out, and was already mentioned in discussing technology integration into Philadelphia schools in Chapter I. Her school was a magnet school. Administrators interviewed about how computing moved into Philadelphia schools noted that the schools that historically received the last integration of technology were the schools where students excelled academically. The assumption appeared to be that these students were able to excel without the additional technology enhancements, so the district should support the poorer performing schools before those that were performing well. The culture in the school was to not include technology since the infrastructure was not there. When Amanda was interviewed in 2006, the school had just received wireless technology, each middle school teacher had received a wireless ready laptop computer, and the school was to receive a small number of smart boards and a mobile computer lab. As more technology was easily available to teachers, one would hope to see more used in this school.

The amount of change that occurred in these teachers was wider than any other group of teachers discussed so far. Using Rogers' stages of adoption (1962, 1972,

1983, 1995, & 2003), all six of these teachers had passed through the knowledge, persuasion and decision stages. They had all decided that they should infuse some level of computing technology into their classroom practices. They were all willing to try out computing in their classrooms, moving toward the confirmation stage when they fully implemented computers usage. Other factors impacted how far they moved into the confirmation stage. Amanda was stymied by having to move around, bringing all her teaching materials with her as she roamed the hallways. Sandra, on the other hand, was so intrigued by what computing brought to her mathematics and science classroom that she went on to explore robotics and teaching students how to refurbish computers. Both of these activities supported her teaching of both science and mathematics.

According to Fullan and Pomfret, (1977) there are five dimensions of change within implementation of an educational innovation: changes in materials, structure, roles and behaviors, knowledge and understanding, and in value internalization. Looking at these six teachers, change happened in relation to these five dimensions. Materials changed in that they received a computer for their classroom. Structure changed for the five teachers who redesigned their classrooms to include greater access for student computing as part of classroom practice. This did not happen for Amanda since she did not have a classroom of her own as the other five did. Did roles and behaviors change? Yes, for the most part, but not for Amanda. Sandra took on a new role within her school to support other teachers, to support the computer teacher, and she taught students how to refurbish computers. Amy supported her students by making her classroom available when they needed to use her computers. Elaine changed her practices to require students to be more responsible for their learning. Sam included more computing time to help his students become better writers. Nancy became a resource for her school, but her classroom practices did not change much. All six of these teachers increased their knowledge and understanding of computing, and all but Amanda increased their knowledge about incorporating computing into classroom activities. Lastly, from their comments and discussion of their classroom practices, one can infer that there had been a change in their perceived value of computing in their teaching. Clearly Sam had changed his practices, as had Sandra.

Using the Concerns Based Adoption Model as a lens for analyzing the impact of participating in CPI for these six teachers, the teachers reached different stages of adoption and different levels of use, with Amanda reaching the lowest levels based on the circumstances in her school. All six were aware of computing in education and wanted more information (stage 2). Five of these teachers reached the consequence stage (stage 4) where they were concerned about how using computers in their classroom practices would impact their students' learning. While Sandra and Sam both spoke about how important it was for them that the CPI program encouraged collaboration between teachers, they were referring to the opportunities to collaborate with other CPI teachers. Sandra offered to support other teachers' uses of computers, but showed no evidence that she was collaborating with other teachers by coordinating activities across classrooms.

As for levels of use, Amanda reached Level 2 (Preparation) by thinking about how to use computers in her teaching, but she just never got to the point where she could incorporate computers into her activities. This is not to disparage Amanda but to note that CPI program outcomes were dependent upon the environments in which the teachers taught. Where 10 to 12 teachers participated and there was a very supportive principal, teachers collaborated across classrooms and across the curriculum. Opportunities were available within the school to support further discussion of how computing power could support student learning. Where there were only a few participating teachers and no evident support by the principal, as in Amanda's case, the impact of CPI on teaching practices was minimal at best. The other five teachers reached Level 3 (mechanical use) by focusing their discussion of using computing in their activities in terms of day-to-day use. None of these teachers were routine users of computing in their practice overall. Again this is not a negative statement, but an acknowledgement that the core curriculum required teachers to be on-topic. If computing solutions were not included in the curriculum framework, the inclination was to use it as a supplement.

Lastly, Fullan (2001) introduced three interactive factors that affect implementation of an educational change—the characteristics of the change, local characteristics, and external factors. Each of these six teachers identified that they were aware they needed to change their behaviors and knowledge related to the use of computers in classrooms. Three were basically non-users and did not want to continue in that vein. Unfortunately, except for Sam who was quite clear that he was interested in how computers could support writing, the goals of these teachers were not clear. Yes, they each wanted to learn more about how to use computers with their students, but for what purpose and to what end? Sandra did not go into the CPI program intent on teaching robotics or how to refurbish computers, but the program inspired her to learn more and more. Amy, Elaine, and Nancy supported their students by allowing them access to computers to work on their projects, but they could have done that without CPI. Each came away from the program learning to manage their classrooms to support computer use by students. They did not know that they would learn this, so they could not have articulated that as a goal. They knew they wanted to use computers with their students, but did not know the importance of classroom management to make it work.

The person most impacted by the local characteristics was Amanda. Sandra, Amy, Elaine, and Nancy all came from schools where there was support for using technology in classroom teaching. With support came access to resources. Sam brought his own technology into his classroom and maintained it himself until he no longer could. Having the infrastructure and administrative support was very important in ensuring what was learned through the CPI program, or any other technology related professional development program as well, would be implemented and maintained over time.

The external factors appeared to be negligible in these cases. While this work was initially supported by IBM and by two Technology Literacy Challenge Fund grants, there was no apparent emphasis by outside agencies compelling these teachers to learn how to integrate technology into their classrooms, or at least none was apparent through the interview process. It was true that the education profession had been moving toward more technology in classrooms since computers were first introduced, but school environments were uneven in making that happen. Many teachers regarded computers as one more burden for them to bear, while a few early-adopters moved easily into using computers regularly. The majority of teachers were in the former category. These six teachers were mixed. Two were already users, Elaine and Sam, while three were non-users. In the end, all but Amanda were implementing computers in their classrooms to support student learning.

References

Docterman, D.A. (1998). *Great teaching in the one-computer classroom*. Watertown, MA: Tom Snyder Productions.

Fullan, M. (2001) *The new meaning of educational change.* Third Edition. New York: Teachers College Press.

Fullan, M., & Pomfret, A. (1977). Research on curriculum and instruction implementation. *Review of Educational Research, 47*(1), 335-397.

Hord, S.M., Rutherford, W.L., Huling-Austin, L., & Hall, G.E. (1987). *Taking charge of change*. Alexandria, VA: ASCD.

Piaget, J. (1972). *The psychology of the child.* New York: Basic Books.

Piaget, J. (1990). *The child's conception of the world.* New York: Littlefield Adams.

Rogers, E.M. (1962, 1972, 1983, 1995, & 2003). *Diffusion of innovations*. Glencoe, IL: The Free Press.

Chapter XIII

Beyond Computers: Grade 8

Technology of the printing press provided a new social and intellectual context for modern schools. The technology of print became the 'defining' technology of schools. A defining technology is a technology that results in fundamental changes in how people see themselves and their world (Norton & Wiburg, 2003, p. 4).

How do you differentiate eighth grade from other grades? Whether the eighth grade is the last grade in an elementary school or the last grade in a middle school, these are the oldest students in the school. When they graduate from eighth grade many are on their way to high school. By eighth grade the students are 13 or 14 years of age, well into the angst years of teenage.

As with seventh graders described in Chapter XII, eighth graders in elementary schools have a homeroom teacher who is either the math and science teacher or the literacy and social studies teacher—or perhaps the special education teacher if they have been diagnosed with a disability and have an Individual Education Plan (IEP) on file. During the school day, they move back and forth between the classrooms as the schedule or IEP dictates, but they will only have these two or three primary teachers. Eighth graders in middle schools may have a homeroom teacher, but that person will only teach one subject or may be a special education or ESL teacher. These students rotate around the school to their specific subject teachers. These two scenarios present very different learning experiences for students, and very different teaching experiences for teachers.

Another aspect of middle school teachers is that they typically teach more than one grade level. Many of the teachers discussed in this chapter taught sixth and seventh graders as well, but discussion focused on eighth grade issues. It is not impossible for a middle school teacher to only teach one grade level, especially in very large schools, but not likely in Philadelphia where middle schools are an endangered species.

This chapter presents seven teachers who all taught in middle school settings. Three of them came from the same middle school in southwestern Philadelphia, two came from another middle school in southwestern Philadelphia, and two came from a different middle school in northwestern Philadelphia. Four of these teachers were special education teachers, meaning they taught more than one subject. One aspect of these stories was how different the impact of CPI was on their classroom practices. This was also true with the seventh grade teachers, and perhaps a phenomenon related to the nature of middle school classrooms. Teachers in middle schools, where students rotate from class to class, function more like high school teachers than elementary school teachers. Their subject matter expertise is what matters most, and the curricula are more focused on higher order thinking skills, rather than on creative expression through drawing, except in art classes.

Table 9 gives the reader an overview of the teachers presented in this chapter. The information provided includes how many years of teaching experience they had

Table 9. Eighth grade teacher summary

Name	Years teaching	Grades in school	Prior computer usage	Overall change	CPI teacher
Ruby	5 years	5 – 8	Novice user	Extensive	VT only
Rose	20 years	5 – 8	Novice user	Very much	VT only
Tanya	6 years	5 – 8	Novice user	Some	VT only
Jerry	26 years	5 – 8	Novice user	Some	VT only
Maddy	26 years	5 – 8	Novice user	Some	VT only
Dara	18 years	5 – 8 to be K – 8	Novice user	Some	VT only
Ella	13 years	5 – 8 to be K – 8	Novice user	Little	VT only

prior to participating in the CPI program, what grades were included in their school, how they characterized their prior use of computers/technology prior to the CPI program, and whether they were a Visiting Teacher or also asked to be a Resident Teacher. This information may help the reader in deciding which teacher story is close to their own.

Ruby

Ruby was in her second career. After a number of years in corporate America, she decided her calling was to be a math and science teacher in a middle school. She had been teaching five years, all at the same middle school where Clara taught sixth grade (see Chapter XI for school demographics), and where Sandra and Amy taught seventh grade (see Chapter XII). During that time, Ruby also took courses toward earning her principal certification, and she earned teaching awards each year she taught. Ruby noted that a previous assistant principal encouraged her to pursue a track to become a principal, given her experience in the business world and her expertise as a teacher (anonymous, personal communication, May 17, 2005).

She was encouraged to participate in CPI by the same assistant principal who encouraged her to pursue the principal certification. This assistant principal encouraged the other teachers at this school as well. Ruby said, "He thought it was important for teachers to take advantage of opportunities to learn more about technology and to earn technology for the school. He's a principal at a school now and really encourages technology in the classroom" (anonymous, personal communication, May 17, 2005). Ruby said she first earned her teaching certificate in the 1970s, when computers were not part of the teacher preparation programs, and she wanted to update her skills. She thought the way she would use computers in the classroom would be different from in the corporate world. She wanted to learn from the people who were putting these curricula together, and, she wanted a computer for her classroom. She admitted to being a little intimidated at first, but her mentoring and the networking with other teachers was very beneficial.

After CPI training, I changed my instructional style to incorporate many different methods. I started using small learning groups, math games to reinforce skills, research on the Internet for science projects, writing assignments, and portfolios of work every six weeks. Students are encouraged to use the concept mapping program Inspiration for planning projects and papers, and to incorporate graphics into their work. Overall, CPI broadened my horizons about lessons and activities. My students do amazing projects! I encourage them to expand their horizons as well, trying to empower them to take charge of their education.

Ruby noted her students mostly did not have computers at home, so she opened up her classroom to help them feel more comfortable with computers. She had three computers and she liked to see them used as often as possible by her students. Like Amy, from the same school, Ruby found that students came to the classroom before school or during free periods to work on projects since she made the computers available to them. She said, "The more comfortable they are, the more they feel in control and independent in their learning."

Each year Ruby had the students do an environmental project working with a local creek conservation project. They created a Web site on the history of the creek pollution and how that was damaging their neighborhood. This project helped them learn about ecology, pollution, and the relationship between the two topics and themselves.

In 2005 the students decided to create a "yellow pages" like resource for kids. The book had over 80 pages of information about different issues teenagers needed to know about, with references to agencies and organizations including Web sites and contact information. Topics ranged from AIDS to the local creek conservation project and to various teen related agencies in the neighborhood. They created a cover for the project, and Ruby was getting it published with the names of the students who contributed listed. Even though these students did not necessarily have computers at home and knew the audience for their book did not have computers at home, they insisted that each entry had to have a Web site for accessing information to be credible.

Ruby took pride in the accomplishments of her students who won awards for their science projects. She may not have changed her practices as much as many critics of technology use in classrooms might want to see, but she started using more constructivist approaches to teaching science and math with small group work and, especially, the local creek conservation project that brought to life biology and quality of life issues in the neighborhood. Ruby credited CPI with giving her access to other teachers from whom she gained the courage to branch out from her normal classroom practice. She brought technology into the neighborhood and helped her students see the connection between the neighborhood and their learning.

Rose

Rose taught special education students. She had taught at a middle school in southwestern Philadelphia for five years, and had taught at a high school for over 20 years before transferring to the middle school. This middle school was the same middle school where Ruby taught science and math to eighth graders.

Rose wanted to participate in CPI to learn how to use a computer, first and foremost, and to get a working computer for her classroom. Rose said she was not shy in noting the need for getting a "working" computer in her classroom (anonymous, personal communication, May 17, 2005).

Not only did I learn how to use a computer, but I learned how to create things I can use with my students. I learned about iMovies, PowerPoint, Web sites for clip art and graphics, and how to navigate and search the Internet in general. We each created a Web page that I have since lost track of.

In spring 2005, Rose had three computers in her classroom, two of which worked at any one time. She had the students rotate onto the computers in pairs. She learned how to do this through visiting her Resident Teacher who showed her how to organize the day to use the computers with different subject areas. Rose said the students in the Resident Teacher's classroom showed her how to teach her students to use PowerPoint. Rose said, "It was so amazing that the students were teaching me, and I was not shy about it." So she taught her students to use PowerPoint and they used it with various projects.

Before CPI, Rose used the computer in her classroom for students to type journal entries or for pre-packaged drill programs or games. After CPI training, she had the students typing their journals and adding digital pictures, clip art or graphics to make it more interesting for them to both do the journaling and view. They did cover pages for projects and for their exit projects they used PowerPoint.

Rose said her CPI teacher "cohort" had sort of a bi-level grouping. There were several members who had minimal to no computer skills, meaning they were learning to turn on the computer and to open applications, and several who had basic skills who needed to learn to use educational software with their students. The cohort was divided into two groups. One group came weekly to gain basic computer skills, while the other group was more independent, doing assignments and coming to training sessions every two weeks. She noted, "We all got together on the two week sessions, which was great for networking and sharing. And we had the online discussion board where we also shared resources and information." All of this had been a boon for her classroom, and also for her master's degree graduate studies projects, something she did not expect.

Again, Rose's classroom practices changed, but not dramatically. Her special education students needed to have a sense of accomplishment. Using the computers to support the students' writing and projects was a way for Rose to empower these students who rarely saw their accomplishments celebrated. CPI gave Rose the opportunity to gain from the Resident Teacher, but also from the other CPI teachers. Sharing with other teachers was a valuable resource for Rose. And most importantly, Rose was able to use the technology in pedagogically appropriate ways to support the curriculum and her students' learning. Seven teachers from her school participated in CPI, five of whom were interviewed for this book. While they did not all teach the same subjects, they supported one another across the school.

Tanya

Tanya taught at a middle school in southwestern Philadelphia, the same school where Shelly taught sixth grade (see Chapter XI for school demographic information) and Nancy taught seventh grade (see Chapter XII). She had taught for only six years, five of which were been at this middle school. She taught mathematics to seventh and eighth grade students, but we mostly talked about her eighth grade classroom.

In the early 21st century, Tanya said she did not use computers with her students at all. She had two older computers in her classroom, but they were not used with her students or within the curriculum. It was not that Tanya did not know how to use computers; she simply did not use them with her students. Tanya said (anonymous, personal communication, June 7, 2006):

I participated in CPI to learn more about technology and to get a [working] computer for my classroom. In the end I learned about PowerPoint, slideshows, Inspiration, and iMovies. I learned quite a lot from the program. I came back to my classroom and taught my students how to troubleshoot computer problems and they, in turn, helped other teachers with their computers. It was a win-win situation all around. We were able to get e-mail accounts for all the children so teachers could contact them outside of school and the children could submit assignments through the e-mail.

In 2006, Tanya's students were completing projects on the computer and using Inspiration to help master concepts. She had a large-screen TV in her classroom with an adaptor to connect a computer so she could display to the class what she was doing on the computer, or a Web site, or even student work. She could use slideshows or CD-ROM materials over the TV to support her students' learning of Algebra. While she had computers in her classroom before, they were older and did not support the Internet or CD-ROM technology. CPI afforded Tanya opportunities she did not have before.

Educational technology critics would say that Tanya did not really change her teaching by incorporating technology into her classroom. Indeed, she remained the "person in charge" rather than becoming a constructivist teacher and helping her students partner in the teaching and learning process (Becker & Ravitz, 1999). However, CPI was not about becoming constructivist teachers as much as it was about incorporating technology in pedagogically appropriate ways to support the curriculum and student learning. Tanya used the large-screen TV to project images that helped her students understand Algebra because this worked better than writing on the chalkboard. Technology brought to her classroom a means to an end by making mathematics more dynamic.

Jerry

Jerry taught special education at the same middle school in southwestern Philadelphia where Tanya taught. She had taught in the district for 26 years, all at this school. She taught sixth, seventh, and eighth grade students, but talked mostly about eighth grade.

In late 1990s and early 21st century, Jerry had a couple of Apple IIE computers in her classroom, ones that no one else wanted. They were only good for typing since most of the software was either missing or out of date. So she did not use them much at all with her students. Then, CPI came along. Jerry said (anonymous, personal communication, June 7, 2006):

I wanted a [modern, working] computer for my classroom. I also wanted to be able to bring up-to-date resources into my classroom for my students. Between the computer programs I received with the new computer and the Internet resources I learned about, all for low achieving students in math and reading, I was able to help my students better than before.

Jerry got one computer for her classroom, but got hooked on what technology could do for her students. She decided to take as many training programs as she could that would provide more resources for her classroom. "I just kept going back!," she said. In 2006 she had three computers in her classroom and she used them for her students' writing skills. Once they wrote a story, report, or journal entry and got it "right,"[1] they were asked to type them up on one of the computers to be reviewed. Jerry noted, "For these [low achieving] children, being able to type up their writing and print it out makes it more real for them. They get a sense of accomplishment that, for their generation, doesn't come from writing with paper and pencil."

Recently, the school gained access to a Web resource called BrainPOP. Jerry would never have considered using the Internet with her students, much less a resource like BrainPOP, before she participated in the CPI program. Now she brings ideas to her principal to be shared with all students in the school. Jerry has become a technology resource for her school, a role that she could never have imagined before her CPI experiences.

Jerry noted that the current principal was supportive, as was Jerry's previous principal. Having that support—someone who sees the connection between technology and student learning—provided teachers with the incentive to try new ideas, new resources, and new learning activities. CPI gave Jerry the comfort with technology needed to become a catalyst for new ideas for using technology in her school. This was something she did not do before because she lacked the self confidence to do so.

Maddy

Maddy had taught in the School District of Philadelphia for 26 years, of which 24 were at the same middle school in southwestern Philadelphia; the same school where Shelly taught sixth grade (see Chapter XI for school demographic information), Nancy taught seventh grade (see Chapter XII), and Tanya and Jerry taught eighth grade. She taught special education, like Jerry, in all grades in the school, but again mostly talked about eighth grade.

In the early 21st century she had no computers in her classroom and in 2006 she had three computers. Maddy wanted to participate in CPI to obtain a working computer for her classroom and to learn more about using technology with middle grade children. Maddy said,

I had a computer at home that my son, who was in the eighth grade, used. I thought if I learned more about computers and got one for my classroom I could transfer work from home to school, and back again (anonymous, personal communication, June 7, 2006).

Maddy reported she had her students do research for projects and that some of the boys could program computers for other teachers. She also had a new science program for the seventh and eighth graders that required using computer resources.

And we learned how to make a Web page, Inspiration and PowerPoint [through CPI]. And we now have access to BrainPOP, which has lots of resources for every grade and subject. They do quizzes on line and the students love it!

Overall, visiting another teacher's classroom was something that was special about CPI, according to Maddy. Teachers learned from other teachers, not just about using computers but about other things as well, and they shared resources with each other both through the CPI program and with each other at the school. In the end there were nine teachers from this middle school who participated in the CPI program and came back to help each other. CPI not only taught teachers computer skills and afforded them time to visit other teachers' classrooms to gain ideas, but provided opportunities for sharing among teachers, also known as peer coaching or mentoring—something that doesn't happen every day in most schools but is quite helpful in supporting teachers' use of technology across the curriculum (Ball, 1996; Berman & McLaughlin, 1975; Fullan & Pomfret, 1977; Guskey, 2000; Joyce & Showers, 2002; Little, 1989; Sharan and Hertz-Lazarowitz, 1980; Wilson and Berne, 1999).

Dara

Dara had taught Special Education at a middle school in northwestern Philadelphia for 18 years when she was interviewed in spring 2005. This was the same school where Ernest taught sixth graders (see Chapter XI for school demographic information). In 2000, when she participated in CPI, she had her own self-contained special education classroom for eighth grade students.

She was not shy about the fact that she participated in the CPI to get a computer for her classroom. She said, "It was the only way to get equipment at that time." Dara said she did not have a computer in her classroom in 2000, and the computer teacher was always encouraging teachers to take the various professional development courses that would provide them with a computer for their classrooms (anonymous, personal communication, April 28, 2005). CPI was attractive because there were a couple of Resident Teachers in her school so she would not have to actually go to another school to observe another teacher's practices.

I learned a lot [from the program]. Today when students have questions I can't answer easily, we go together onto the computers to do an Internet search for information. I use this as an opportunity to reinforce that it's okay to not always have the answers because there are easily reachable resources where the information can be attained. For my special education students, often the problem is not the fact that they don't find the information, but the problem is not being able to read and determine appropriate information.

Dara still used the system she learned to cycle students on and off the computer so everyone got a turn. She indicated she learned a lot through CPI, but without continued practice she had not retained all that she learned. One example she cited was the Web page she created. She had not kept it up since 2000 and did not remember how to do so.

What was interesting about Dara was her desire to bring more resources into her classroom to help her special education students. Special education students rarely have as successful a day as compared with the other students, especially in middle school. Having the technology available for the students to work on projects individually or in pairs allowed Dara to give her students opportunities to be proud of their accomplishments, similar to Rose. While that may not seem valuable for most students, it is extremely valuable for students in special education.

Ella

Ella had taught at the same middle school in northwestern Philadelphia for 13 years. She taught math and science to all grades, but we mostly talked about her eighth grade class. Her school was the same school where Ernest taught sixth grade (see Chapter XI) and Dara taught special education. Before she participated in CPI, Ella said she did not use computers with her students, and indicated she did not see any reason to do so (anonymous, personal communication, April 28, 2005).

I wanted to participate in CPI because my only use of computers in 2000 was e-mail and I wanted to be more computer literate. As the other teachers here who went to CPI, I was influenced by the computer teacher who strongly encouraged me to participate. He was always pushing us to get out there and take the tech courses, especially if we could get a computer for our classroom. Face it, the school budget did not afford buying computers for the classrooms. The students took the computer class, so why did we need computers in the classrooms? But the computer teacher thought differently, and he pushed and pushed us!

Ella said she learned a lot from her CPI experiences. But like Dara, she noted that when skills are not used for a period of time, they tended to recede into the background and to be easily forgotten, especially computers skills. This phenomenon is similar to what Molly described in Chapter XI. In spring 2005, Ella used the computer to search the Internet, doing research for her teaching and for school district work. "I would like to learn more, but I do feel much more confident in my use of computers."

Ella's story is short and to the point, and her classroom practice did not change as significantly as other teachers. In fact, she was only using the CPI computer to search the Internet for resources she could use with her students. Saying that is not to disparage Ella for not exploring more. The reader should remember that, like Ernest from Chapter XI, Ella did not see the point of computers in her classroom. She wanted to be more computer literate for her own sake to support her teaching. While she indicated that computers helped her students, she noted she would probably see even greater accomplishments over time.

Summary

These seven short stories show how change happened in different ways for each of these teachers. They were from three middle schools, two in southwestern Philadelphia and one in northwestern Philadelphia. Four of these teachers were special education teachers, and three were mathematics or mathematics and science teachers. All of them participated to get a working computer for their classroom and all but one, Ella, used the computers in their classrooms to support learning through having students create or complete projects using Inspiration and PowerPoint, in addition to searching for information on the Internet.

Dara and Maddy used the computers they had to empower their special education students to see that they could succeed. For Ruby, the computers in her classroom supported the environmental project she had her students do each year. For Tanya, the computer with the large-screen TV allowed her to project what she was doing on the computer to all her students, helping them to understand Algebra better.

For most of these teachers, CPI was about sharing. Sharing could be with other CPI participants during the training sessions, or through the online Learning Village Web site, or with each other back in their schools when there were several teachers from the same school. This was definitely the case for Ruby, Rose, Tanya, Jerry, and Maddy. Sharing of resources within the school gave support back in the home environment that many teachers so badly needed to continue moving forward to incorporate computers into classroom practices. When one teacher found a resource they thought was valuable to the entire school (e.g., BrainPOP), they shared readily. They may have done so prior to CPI, but not about computer resources. Now they were empowered to be resources for each other around computers.

Analysis

As noted previously, change does not happen easily or in a vacuum, and for these seven teachers change did not happen equally. Each of these teachers taught in a middle school setting. They each had their own stand alone classroom where they either taught mathematics and/or science, or they taught special education students where they taught a number of subjects, including writing. They also came from schools where, as they noted, the school budget did not accommodate purchasing computers for classrooms. While there were a few computers in classrooms, these tended to be older and not Internet compatible. Thus, these teachers found the CPI program an option for getting newer, Internet ready computers for their classrooms.

Using Rogers' diffusion of innovations theory (1962, 1972, 1983, 1995, 2003), all of these teachers came into the CPI program with an awareness of computing. Only two of these teachers, Maddy and Rose, considered themselves as non-users before the CPI experience. Ruby, Tanya, and Jerry indicated they did not use the computers they had with their students because they were old computers and either had old software or only games on them. All of these teachers indicated an interest in becoming more computer literate, with Dara and Ella having to be persuaded by the computer teacher at their school and Ruby and Rose being persuaded by the assistant principal at their school. Having someone they respected and trusted be the motivating force made it even more compelling for these teachers to participate in CPI.

All of these teachers moved through the decision and implementation stages and decided that computers could benefit their classrooms and students, with Jerry wanting to know more. Only one teacher did not integrate computers into her classroom. Ella noted that she wanted to learn more for her own use of computing rather than how it would support her students. She consciously chose not to fully implement computers into her teaching because she was not convinced it was necessary for her classroom. Zhao and Cziko (2001) would argue that some teachers choose not to implement computing into their classrooms because the training did not convince them otherwise. And it is quite possible, as with Ernest, if the computer teacher had not pushed Ella constantly to take technology classes, she might have been quite happy using only e-mail.

The Concerns Based Adoption Model (CBAM) (Hord, Rutherford, Huling-Austin, & Hall, 1987) provided a different set of lenses for analyzing how these teachers adopted computing into their classrooms, or not. As for the seven stages of adoption, all seven teachers were aware of computers and wanted more information. All but Ella moved through stages 2 (personal) and 3 (management) to being concerned about how computers in their classrooms would affect their students. Ruby changed her teaching practices to incorporate computers into projects. Tanya used the large-screen TV connected to her computer to support her students learning. Jerry used her computers to support her students' writing, and Maddy and Rose used their computers to support projects with PowerPoint and Inspiration. Only Ella never moved past the second stage about how the computer could help her.

Looking at the levels of use from CBAM, all seven of these teachers moved into level 3, or mechanical use of the computers. All but Ella reached level 4A, or routine use, but no further. None of these teachers talked about varying the use of computers based on the needs of students. Ruby, Rose, Tanya, Jerry, Maddy, and Dara all talked about how they were using computers to support their students' learning, but through activities that were mostly whole class and not individualized by student. It would have been nice if these teachers had moved into individualizing activities for their students, but it is important to reiterate that the literacy and mathematics curricula changed shortly after these teachers completed the CPI program. And as mentioned before, the new curricula were more demanding by being relatively scripted and fast paced. Computer solutions for teaching the new curricula were not included, so any use of computers was in addition to that required.

Fullan (2001) suggested three interactive factors that affect implementation of an educational change in a classroom—the characteristics of the change, local characteristics, and external factors. All but Ella were convinced of a need for change in teaching practices to include computers. The goals and means for change were clear—take the CPI professional development program to learn how to infuse computers into classrooms and to try it out with their students. Overall, change did not need to be dramatic. On the contrary, getting students to use computers to support various projects was better than no use at all, and might provide opportunities to try new ways of teaching familiar subjects. Ruby had her students create a Web site for an ongoing project, while other eighth grade teachers incorporated the computers into teaching mathematics or writing. These applications of computers in classroom activities turned out to be practical changes.

The local characteristics of these schools proved to be supportive for change. For Ruby and Rose the assistant principal encouraged them to learn more about using computers in the classroom. And for Ruby, the same assistant principal had encouraged her to become a principal. No doubt she would take the example of this assistant principal with her when she became a principal, likewise encouraging her teachers to explore how computers could support their students' learning. Dara and Ella had the support of the computer teacher, along with the other 13 teachers who participated in CPI. Tanya, Jerry and Maddy did not mention anyone specifically in particular at their school who encouraged them to incorporate computers in their practices, but they indicated that the principal encouraged the use of computers in support of classroom practices. All teachers, except Dara and Ella who were from the same school, indicated sharing with other teachers and how important that was for them.

The situation at the school where Dara and Ella taught was of interest in that there were two Resident Teachers at the school and 15 teachers on record for having taken the CPI professional development program. Of those 15 teachers, 11 visited one of the two local Resident Teachers. There had been changes in leadership in the school between 2003 and 2005, and the school was targeted to be reduced a grade level in fall 2005. These changes in the environment could explain the lack of overall social support between the 15 teachers. The reader should also note that none of the three teachers interviewed from this school participated in CPI on their own volition, but at the repeated request of the computer teacher. They did not all teach the same subjects, but CPI was not about teaching a particular subject as much as how to manage the classroom to make use of resources available to support teaching and learning. As surmised by Marla in Chapter V, teachers who participated in CPI because someone else wanted them to do so were less likely to be satisfied in the end, as opposed to those who chose to participate.

As for external factors, which Fullan indicated as being the government or other agencies, none of these teachers spoke about such outside factors. And none of them mentioned a perception that it was "time" for them to learn more or that there was a push from the profession to learn more. The closest to this was Dara and Ella talking about the computer teacher at their school who pushed every teacher in the school to use computers and to become a 21st century teacher, as noted by Ernest in Chapter XI.

Lastly, from the Apple Classrooms of Tomorrow (ACOT) studies, five stages of change were documented for integrating computers into classrooms (Fisher, Dwyer, & Yocam, 1996; Gordon, 1996). Only two of these teachers came into the CPI program as self-described non-users of computing. All of them had moved through the Adoption Stage and were willing to use a computer. For the most part, they used their computers to support familiar teaching activities. Even Ella used her computer to search the Internet for resources she could use in her teaching. Tanya, Rose, Jerry, Maddy, and Dara all had reached the adaptation stage, where they were using their computers to support the curriculum. Jerry noted how student use of the computers to write was improving their writing skills. Rose indicated how having her students create projects using PowerPoint had improved their projects as well. Only Ruby had reached the appropriation stage of change where she encouraged her students to use the computers to be creative in their learning. None of the other eighth grade teachers interviewed were as supportive and encouraging as Ruby was regarding exploring how the computers could expand the opportunities for their students.

Overall, change was evident in each of these teachers. While Ella did not embrace computers for her students' learning of mathematics and science in the classroom, she used the computer to find resources for her teaching. She did not come into the program willingly, but to get the computer teacher to stop pushing her to take technology training; thus, she was not fully invested in the process from the beginning. Dara did not change significantly either, but was willing to encourage her students to use the computers to find information. She was very clear that her concern with her students was not that they could not find appropriate information, but that they could not read the information to know what it said. Tanya learned how to use her computer with a large-screen TV which she then used to project software that supported her students learning Algebra. And Jerry, Maddy, and Rose used their computers to support students in completing projects and writing. Ruby made the greatest change in her practices. She described the change as moving from "the person always in charge" to encouraging students to work in groups and to take control of their learning through technology.

Were any of these changes large? The short answer was no. However, in the long run, these were teachers who were not using computers to support their teaching at all. No one ever said change needed to be extensive. In fact, the process begins for most teachers by exploring how computers can support teaching and learning, especially when provided with the opportunity to observe other teachers modeling sound pedagogy and using computers with their students. But the teacher must come to the training ready and willing to embrace computing as a viable tool for their classroom. Jerry did so and came away wanting to know more, and thus became a resource for her students and the school in general. Ella did not and her only change was to support her own activities, not that of her students. Even so, it could be argued that her students gained from her own exploration of options to support her teaching.

Conclusion of Section IV

As with the teachers from grades Kindergarten through five, these teachers were from different parts of Philadelphia. Most of them were from middle school environments that were much different from elementary schools. While getting a working computer for their classroom was important, supporting their students' learning was of equal importance. Not as many of these middle grade teachers identified themselves as "computer illiterate," but most did not use computers with their students before their CPI experiences. The change in their classroom practices was not as great as that for the earlier grade teachers, which may be due to the fact that they were from middle schools. In middle schools teachers are subject specialists rather than generalists. Students cycle through their schedules with approximately 50 minutes per class period. Infusing computers into a 50-minute class period is quite a challenge. The special education teachers' students were with them for longer periods of time, and computers were often used to support reading and writing.

The middle school environment could have been a major part of the equation as to whether they used computers or not. Nationally, the fate of middle schools is in debate. Why, in the end, the teachers who were from middle school grades were less likely to fully embrace the contribution of computers in their classrooms cannot be answered from this work. As noted by Anfara and Buehler (2005), not all middle schools actually implemented the middle school concept as explicated fully in the *Turning points* (Carnegie Council on Adolescent Development, 1989), *Turning points 2000* (Jackson and Davis, 2000), and *This we believe* (National Middle School Association, 2003). Whether this was the case for these teachers and their schools could be the topic of another project and book.

In the end, almost all of these teachers were using computers to support their teaching in one form or another. While there were a few teachers whose school environment precluded easy use of computers (i.e., Amanda), the others were able to find meaningful ways to incorporate computing into their daily activities. Two specific barriers mentioned were the lack of printing capabilities in the classroom, making it difficult for students to print their work, and the core curricula for the district which were tightly scripted without inclusion of computer activities.

References

Anfara, V. A., & Buehler, A. (2005). Grade configuration and the education of young adolescents. *Middle School Journal*, 37(1), 53-59.

Ball, D. L. (1996). Teacher learning and the mathematics reforms: What we think we know and what we need to learn. *Phi Delta Kappan, 77*(7), 500-508.

Becker, H. J., & Ravitz, J. (1999). The influence of computer and Internet use on teachers' pedagogical practices and perceptions. *Journal of Research on Computing in Education, 31*(4), 356-384.

Berman, P., & McLaughlin, M. W. (1978). *Federal programs supporting educational change, Volume VIII: Implementing and sustaining innovations*. Santa Monica, CA: Rand.

Carnegie Council on Adolescent Development. (1989). *Turning points: Preparing youth for the 21st century*. New York: Carnegie Corporation.

Fisher, C., Dwyer, D. C., & Yocam, K. (Eds.) (1996). *Education and technology: Reflections on computing in classrooms.* San Francisco: Jossey-Bass Publishers and Apple Press.

Fullan, M. (2001). *The new meaning of educational change.* Third Edition. New York: Teachers College Press.

Fullan, M., & Pomfret, A. (1977). Research on curriculum and instruction implementation. *Review of Educational Research, 47*(1), 335-397.

Gordon, A. R. (1996). *Computers in the classroom: How teachers and students are using technology to transform learning*. San Francisco: Jossey-Bass Publishers and Apple Press.

Guskey, T. R. (2000). *Evaluating professional development*. Thousand Oaks, CA: Corwin Press.

Hord, S.M., Rutherford, W.L., Huling-Autsin, L., & Hall, G.E. (1987). *Taking charge of change*. Alexandria, VA: ASCD.

Jackson, D., & Davis, G. A. (2000). *Turning points 2000: Educating adolescents in the 21st century*. New York & Westerville, OH: Teachers College Press & National Middle School Association.

Joyce, B., & Showers, B. (2002). *Student achievement through staff development*. Third edition. Alexandria, VA: ASCD.

Little, J. W. (1989). District policy choices and teachers professional development opportunities. *Educational Evaluation and Policy Analysis, 11*(2), 165-179.

National Middle School Association. (2003). *This we believe: Successful schools for young adolescents*. Westerville, OH.

Norton, P., & Wiburg, K. M. (2003). *Teaching with technology: Designing opportunities to learn*. Second edition. Belmont, CA: Wadsworth/Thomson Learning.

Rogers, E. M. (1962, 1972, 1983, 1995, & 2003). *Diffusion of innovations*. Glencoe, IL: The Free Press.

Sharan, S., & Hertz-Lazarowitz, R. (1980). A group investigation method of co-operative learning in the classroom. In S. Sharon, P. Hare, C. Webb, & R. Hertz-Lawarowitz (Eds), *Cooperation in education*. Provo, UT: Brigham Young University Press, pp. 14-46.

Wilson, S. M., & Berne, J. (1999). Teacher learning and the acquisition of professional knowledge: An examination of research on contemporary professional development. In A. Iran-Nejad, & P.D. Pearson (Eds), *Review of Research in Education*, Volume 24, 173-209.

Zhao, Y., & Cziko, G. A. (2001). Teacher adoption of technology: A perceptual control theory perspective. *Journal of Technology and Teacher Education, 9*(1), 5-30.

Endnote

1 Getting it "right" translated into having full sentences and paragraphs, having proper punctuation, and having completed the assignment in a timely fashion.

Section V

Conclusion and Professional Development Model

This last chapter in this book summarizes the book as a whole and provides a conclusion. The Continuous Practice Improvement (CPI) staff development program was studied for three years before it was eliminated due to cost factors, and an additional three years were required to gain the teacher stories found in this book. The major cost of the program was the substitute teachers, which as noted in Chapter IV were difficult to find for grades above fourth grade. The teachers interviewed for this book had a wide range of experience as teachers, from as low as three years to as high as nearly 40 years. They also were at different points after participating in the CPI program ranging from three to seven years. Most staff development programs do not have an impact that spans past two to three years post training, but CPI was still having an impact for many of these teachers seven years afterwards.

Chapter XIV also brings together the theories used as lenses to study how CPI impacted teaching practices of these 57 teachers. From the theories and the outcomes of the overall study of CPI over the six years, a model for how staff development about infusing computers into classroom practices is presented. This model is the culmination of the research on CPI and represents the cumulative outcomes for these 57 teachers in this school district.

Chapter XIV

Effecting Change in the Classroom Through Professional Development

Computers are not an end in themselves. The goal of technology integration into the classroom and curriculum is not to expose students to computers and the Internet. Technology, by definition, is a tool. That tool should be serving a greater purpose, and what that purpose is for you may be very different from what it is for the school down the street (Dockterman, 1998, p. 21).

Over the last 10 to 15 years, a vast amount of money has been invested in computer technology for public education, including establishing or upgrading computer labs, school and classroom connectivity to the Internet and wide area networks, and providing at least one Internet-ready computer per classroom. The Department of Education's Education Rate, or E-Rate, initiative[1] has made access to computing and Internet connectivity easier, reducing the ratio of students to computers to the 5 to 1 ratio that "many experts consider … a reasonable level for the effective use of computers within the schools" (President's Committee of Advisors on Science and Technology 1997, p. 14, as cited in Cattagni & Farris, 2001). In addition, we have seen the introduction of mobile computer labs in schools and, more recently, smartboard technology in classrooms. Each innovation has represented a decision by a school district to use scarce resources to bring computing power into classrooms to affect teaching and learning. However, just because tools are present in classrooms does not mean teachers know how to make the most of those tools, so professional development opportunities have been created to help the teachers learn how to use the tools. Not all professional development programs have been equal in helping teachers make the most of new technologies in their classrooms.

Organizations, conferences, and journals have been dedicated to presenting research on the impact of computers on learning and offering best practices for teachers and administrators. Examples of implementation programs, professional development models, and classroom management tips have been discussed in all venues, with the intent to influence practice in school districts and classrooms. However, the real tests of the technology impacts in classrooms and on student learning are (1) whether the teacher knows and demonstrates how to use computers in pedagogically appropriate ways, and (2) whether student achievement is enhanced as a result.

The history of classroom computing is brief. Early uses of computers were mostly for drill-and-practice of skills, and/or for learning programming languages like LOGO or BASIC. Computer training for teachers included how to use a computer for computer-assisted instruction and/or how to use computers to create worksheets, classroom banners, or acetate slides to use with the overhead projector. Other than drill-and-practice, students used computers in creative ways through paint and draw programs, but not much else until the late 1980s when computer productivity tools came into schools. Then some enterprising teachers saw the power of word processing software in helping students with the writing process. Overall, the computer has been an add-on to the classroom, not a must-have for teachers.

The fact that most teachers have only used classroom computers with their students for games or drill-and-practice to reinforce concepts, or as a reward for good behavior, is disappointing, but not news (Cuban, 2001). Most public school teachers were not taught in elementary, secondary or post-secondary classrooms that used computers at all, or they were used only to reward good behavior. If they had computers in their public school education it was most likely in a computer lab divorced from the curriculum in the rest of the school. Therefore, they cannot have a mental paradigm for infusing computing seamlessly into classroom activities. In addition, since the NCLB legislation (U.S. Department of Education, 2001), several school districts have tightly controlled math and literacy curricula, leaving teachers very little latitude for incorporating computer solutions to support their students' learning.

Noted in Chapter I, as computers became more powerful and available in classrooms, as software applications evolved beyond electronic programmed texts, and as the Internet became available to schools, the potential use of computers began to move away from passive learning modalities to active learning practices. Unfortunately, other than the Apple Computer experiments in the late 1980s and early 1990s, professional development for teachers on uses of computers in classrooms continued to follow the traditional models of general teacher professional development.

Another key factor for teachers in schools is what Lee Shulman called the "pedagogical solitude" that teachers experience (Shulman, 1993). In this article he was addressing post-secondary teachers, but his comments are equally applicable to K-12 teachers. "We close the classroom door and experience pedagogical solitude" (Shulman, 1993, p. 6). How many teachers can identify with the experience of being isolated in their classroom with their students? Rarely do teachers have the time or opportunities to observe each other's teaching styles and methods. Rarely do teachers take the time to share with others some new teaching resource they just found and how it worked with their children, or not. In fact the methods that do not work are even more rarely noted. But even the ACOT studies spoke to the fact that when teachers observed other teachers using technology successfully with their students, the Visiting Teachers came away with a sense of empowerment—they, too, could make this happen with their students (Fisher, Dwyer, & Yocam, 1996) Observing peers successfully infuse computers into daily teaching and learning activities proved an essential component for the success of the CPI program, as demonstrated by the teachers presented in this book.

Change Theories

In Chapters V through XIII, several lenses were used to assess whether change in behaviors had occurred and at what levels. These lenses were Rogers' diffusion of innovation theory (1962, 1972, 1983, 1995, 2003), Berman and McLaughlin's perspective on successful implementation of an innovation in schools (1978), Fullan and Pomfret's dimensions of implementation of change (1977), the Concerns Based Adoption Model (Hord, Rutherford, Huling-Austin, & Hall, 1987), the stages of technology adoption from the ACOT studies (Dwyer, Ringstaff & Sandholtz, 1990; Fisher, et al, 1996; Gordon, 1996), and three interactive factors that impact implementation of educational innovations proposed by Fullan in 2001. Each of these will be briefly reviewed and then the impact of the CPI professional development program on teachers' classroom practices will be discussed. One factor to consider is the level of change in teacher classroom practices. For this book, the measure of change was not extensive change, but that teachers embraced the use of computers as part of daily practice, even if humble in nature.

Rogers' diffusion of innovation theory (1962, 1972, 1983, 1995, 2003) was not specific to education or technology, but quite applicable in understanding how adoption occurs or not. The five stages of adoption included awareness of the innovation, interest in learning more about the innovation, evaluation of the innovation, trying the innovation out, and adoption or rejection given information gained through this process. In addition, he noted five criteria the innovation must meet for adoption to be likely—considered better than current practice, compatible with school/classroom culture, simple and easy to understand, easy to test out, and observable in another teacher's classroom. The intersection between the stages of adoption and the criteria allow the reader to determine whether the innovation was adopted, thereby signifying change in practice.

Berman and McLaughlin (1978) reported that well-implemented reforms resulting in changes in classroom practices had a strong training component with practical workshops, locally available technical assistance, and effective follow-up support to address the teachers' specific problems in implementing the strategy in their classrooms: "Classroom changes were most apparent where the projects provided teachers with some form of classroom support, which complemented the training and/or materials development phases of the project" (Berman & McLaughlin, 1978, p. 45). These findings are compatible with those in the ACOT studies where the resources were provided by the supporting agency, not by the school districts involved.

A third theory considered was that of Fullan and Pomfret (1977). In their review of research on curriculum and instruction implementation, they offered five dimensions of implantation of change in practice: changes in materials, structure, role/behavior, knowledge and understanding, and value internalization. These findings resonate with those reported in the ACOT studies, where teachers changed classroom practice only over time with regular access to computers in their classrooms and ongoing technical and teaching support (Dwyer, et al, 1990).

Hord and colleagues (1987) had studied the adoption of educational changes in schools for many years when they put forth their Concerns Based Adoption Model. In this model, two frames are used to understand adoption of an innovation—stages of adoption and levels of use. The stages of adoption range on a continuum from an awareness of the innovation but no interest, to concern about how use of the innovation will impact the teacher personally, to concern about how the innovation will support the students' learning, to collaborating with other teachers with the innovation, and lastly to refocusing the use of the innovation to further improve outcomes. The levels of use range on a continuum from non-use to learning about and getting ready to use the innovation, to focusing on how to use the innovation daily, to routine use without special preparation, to varying use based on student needs, to collaborating with colleagues, and lastly to re-envisioning the use of the innovation within the school.

While not specifically a "change theory" per se, the ACOT studies chronicled how change occurred over time in classroom practice for using computers where there was extensive technology and technical support available. These were entry stage with no knowledge of computers in classrooms, adoption stage where they were willing to learn, adaptation stage where the computers were used to support the curriculum, appropriation stage where computers were used in project-based learning, and the invention stage where teachers used computers in innovative ways (Fisher et al., 1996; Gordon, 1996) One major finding from the ACOT studies was the importance of "time" for change in teacher practice to occur. Teachers must become comfortable with their own knowledge with computers before they can use them with their students—much less consider non-conventional uses of computers within the curriculum.

Lastly, Fullan (2001) suggested three factors to consider when assessing implementation of an educational innovation or change. These included the characteristics of the change—is it needed, is it easy to implement, is it high quality and practical, the local school characteristics—the principal, the teachers, the culture in the school supporting change, and any external influences—expectations by the district staff, professional communities, or federal mandates. While these factors are not listed in any specific order of influence, they confirm previous theoretical lenses, and are all used in analysis of these 57 teachers.

CPI and Change in Teacher Behaviors

The teachers presented in this book all had different reasons and rationales for participating in the CPI professional development program. Some thought it was time they came into the 21st century in using computers with their students. Many came wanting the "working" computer and printer and then became hooked on how computers could engage their students in learning. Some came kicking and screaming into the program and came away with less enthusiasm for how technology could support their students. But change did happen for these teachers. Using the six lenses from change theory, and the experiences from the ACOT studies, the factors that impacted these teachers will be analyzed.

Rogers, Berman and McLaughlin, CBAM, Fullan and Pomfret, the ACOT studies, and Fullan all note that the nature of the innovation or educational change is important to assess. The easier the change/innovation is to understand, and the better it fits with current practices/values/classroom culture, the more likely there will be change in practices. Making the innovation or change easy to understand was one of the assets of CPI as described by the teachers. While there were expectations in the program that participating teachers would learn more about computers and how computers could support their classrooms, there was no illusion that all CPI teachers would graduate from the program with the same set of skills and knowledge. The program director made it clear that each teacher would learn "at their own pace" and gain skills that met their needs, not the needs of the program. In the end, many of these teachers learned from the Resident Teachers how to establish student computer rotation systems and from CPI how to create presentations and develop Web sites. They came away from their rotation visits with a wide variety of tangible materials, from the CD-ROM of information created by one Resident Teacher to materials printed out in the classroom as described by Marla (see Chapter V). The sophistication and extent of these outcomes depended on the individual CPI Visiting Teacher. Going beyond the classroom/laboratory experience to seeing another teacher's classroom in action made changing a teacher's own classroom practices seem possible. As several teachers said, they came away from the total set of experiences with endless possibilities for how computers could support their students' learning. For those who embraced the change, computers in their classroom practices became something that fit into their new view of what classroom practices should be.

Berman and McLaughlin (1978) noted that for change to occur in classroom practices there needed to be a strong training component. No one would argue otherwise about CPI. CPI teachers all described the training as gaining knowledge about computers in general and how to access, evaluate, and use Internet resources with their students. Consider the stories of Gina (see Chapter VII) and June (see Chapter IX). Gina had no use for computers in her classroom. She noted she used computers with her seventh graders when she taught seventh grade, but only through the computer class. Through her CPI experiences Gina gained a respect for what computers could bring to the early grade classrooms to support students and became an advocate for computer use in her school, as well as within the CPI program as a Resident Teacher. June thought she was going to break the computer, or it would break her. After her CPI experiences she was eager for her students to embrace the wealth of educational opportunities available through the Internet to support learning. These were two extreme stories, but indicative of the outside power of the CPI program in changing attitudes and behaviors.

Almost all of these teachers described their experiences with the Resident Teachers as informative and educational, with the Visiting Teachers returning to their home classrooms full of ideas for the endless possibilities afforded through the computers and the Internet for their students. The teachers who did not find the training strong were those who came into the program with extensive background in computing, like Elaine from Chapter XII who was a computer major in college. Teachers who returned to their classrooms without change in practice were more likely those who participated only because another teacher cajoled them into participating, without a personal need to do so. The majority of the CPI teachers described themselves as non-users or novice users of computers, and the training as helping them gain confidence in their ability to successfully use computers with their students.

Time was an important factor in Rogers' diffusion of innovations theory and the experiences from the ACOT studies. Time to practice new skills, to develop an interest in using computer resources with their students, to see students using computers in the Resident Teacher's classroom, and to determine whether computers would support their classrooms was an essential component of the CPI program. From the training component, teachers were afforded a laptop loaner computer to use at home or at school to practice the exercises from the Saturday sessions and to participate in the online discussion group through Learning Village. The most important time was when they left their home classroom to visit the Resident Teacher for one or two days. Having the time to visit another teacher, the quality of time that, as one teacher described it, was not scripted but real, made a significant difference for these teachers. It is fine to describe how an ideal classroom works, but seeing is believing. In some cases, the students in the Resident Teacher's classroom taught the Visiting Teacher how to make the system work through example. Having another teacher share what worked and did not work in their classroom empowered CPI Visiting Teachers to explore options in their own home classrooms. Teachers often came back to their classrooms with a better idea of how to establish students as "experts" who could support the use of computers in their classrooms.

Support in using an innovation in the classroom was mentioned by Rogers, Berman and McLaughlin, and from the ACOT projects. Support through the CPI program took a two-dimensional approach. Social support in the school was demonstrated by several teachers where a number of teachers from the same school participated together in the program—demonstrating the power of peer coaching or mentoring. (Ball, 1996; Berman & McLaughlin, 1978, Fullan & Pomfret, 1977; Guskey, 2000; Joyce & Showers, 2002; Little, 1989; Sharan & Hertz-Lazarowitz, 1980; Wilson & Berne, 1999) For Marta and Nancy in their school, as well as Darrell, Edie, and Dee in their school, having the same lunch period gave them time to discuss ideas, share resources, and support each other in their quest to infuse computer resources into their classrooms. For Patrice and Mary Jo, having someone with whom to ride to and from school allowed them to share in quality time as well. In other situations, teachers supported each other through grade group meetings, or by carving out time before or after school to meet and talk about what they had learned, what worked for them, and how they were changing their classrooms. Where there was less of a social support network, changes in classroom practices were not as clear or sustained over time, as demonstrated by Maya who was easily distracted by a new operating system and Amanda who was limited by the layout of her school and not having a dedicated classroom of her own. Saying such does not disparage or denigrate the experiences for these teachers, but acknowledges the circumstances in which they taught.

One aspect of support that had not been expected, but was predicted in Chapter I, was the importance of the principal in supporting the teachers' participation in the CPI program. One example was presented by Patrice and Mary Jo who described their principal as supportive of any teacher who wanted to participate in the CPI program. Jo and Gina also noted how their principal saw the power of the digital camera for teaching and bought more cameras for the school. In addition, they noted that their principal did not prevent a CPI Visiting Teacher from visiting their assigned Resident Teacher when no substitute teacher was available. She did what was needed to ensure the Visiting Teacher's classroom was provided support while the Visiting Teacher was away. Similarly, Mickie, Geri, Kim, Margaret, Becca, and Elaine noted that their principal sent someone to visit the classroom of the Visiting Teacher while they were on rotation with the Resident Teacher, just to be sure all was well. Where strong support from the principal was manifest, more teachers participated. Where less support existed, fewer teachers participated. Teachers with a principal who understood how computers could transform a classroom from a passive learning environment into one that supported active learning were given the opportunity, through CPI, to remake their classrooms to reflect the power technology could bring to the teaching and learning process. However, having a supportive principal, and many colleagues participating, did not guarantee all teachers embraced computers into their classrooms—as evidenced by Ernest in Chapter XI.

In all cases, technology support back in the classroom was important. Knowing that the new computer was set up properly and that the teachers could contact the CPI director any time for advice was reassuring. They knew they were not alone in their classrooms with technology. Since the use of the computer(s) was not the only solution to teaching and learning in their classrooms, if there were a glitch, as could be the case in those schools where the infrastructure did not support multiple users in a classroom, these teachers knew they could move on seamlessly without being concerned that they were seen by the students as lacking appropriate skills. Teachers cannot control, or solve, the infrastructure issues faced in older schools and classrooms. Having examples from the Resident Teachers, and the CPI training for how to manage the classroom when the technology is not working well, empowered these teachers to not be afraid of taking risks with their children.

Fullan and Pomfret (1977) suggest that the ultimate indication of successful implementation of an innovation is a change in values, including changes in classroom structures, roles and behaviors, knowledge and understanding, and uses of the innovation in the classroom. In the case of all but a few of these teachers, there were changes in classroom structures (e.g., establishing a computer center as a learning area in the classroom), behaviors (e.g., using computers with their students when none existed before), roles (e.g., identifying student 'experts' to help students when problems arose), knowledge and understanding from the training and Resident Teachers, and the value of technology in classrooms. Only a few teachers expressed that their use of computers in their teaching had virtually not changed. These were teachers who were not really interested in the program because they thought it would benefit their teaching, but to stop another teacher from pushing them toward technology. The program simply could not get beyond the barriers of genuine disinterest in computers for supporting their work. And then there were examples where the teachers did not see the point of computers at all who became converted to the power of the computer to inspire their students learning.

From the CBAM stages of adoption, these teachers entered the CPI program aware of computers in classrooms and, most of them were interested in learning more. All but a very few entered the stage of concern about how the computer would support their students' learning, and participants from schools with a large number of CPI teachers reached the stage of collaboration with other teachers with the innovation. There were instances where the infrastructure of the building or the new curriculum were noted as impediments to continued use of computers with their students, which was true for Rhoda (see Chapter V) and Darrell, Edie, and Dee (see Chapter IX). The level of use prior to coming into the CPI program, ranged from non-use to having used computers since the early 1980s. After participation in CPI, computer usage level in classrooms ranged between focusing on how to use the innovation daily to routine use without special preparation. None of these teachers reached the level of use where they differentiated uses of computer options with their students, or at least none described such activities. The greatest change in behaviors for these teachers was moving from allowing older computers to gather dust in the back of the classroom to having students create presentations or complete projects based on the curriculum—demonstrating the power of the computer to help these students express or demonstrate their learning.

In the end, all of these components were important to ensure sustained change in practice. Each was important in its own right, but together they provided a powerful framework to ensure sustained change in classroom practices. If school districts spend scarce resources to put computers in classrooms and to provide professional development to help teachers transform their classrooms to take advantage of what computers can bring to the learning environment, there needs to be some assurance that the investments are worthwhile. This framework is introduced as one that combines the best aspects from Rogers' theory, the findings from Berman and McLaughlin and Fullan and Pomfret, the CBAM and ACOT studies and Fullan, along with the important ingredient of the principal, to support successful technology oriented professional development of teachers as exemplified by the CPI professional development model.

Discussion

The impact of CPI on practice often appeared as "slight" changes in practices to include typing of written materials (e.g., journal entries, stories) and Internet searches, and sometimes more extensive practices where students completed projects using PowerPoint or created Web pages to demonstrate their knowledge. According to Rogers' theory of diffusion of an innovation and the ACOT studies, "significant" change in teaching with technology takes years to accomplish, not weeks or months.

These CPI teachers reported increases in knowledge and skill in using a computer for personal and teaching purposes, even if they had not practiced those skills since the training was completed between three and seven years before being interviewed. They gained confidence in using computer technology and acquired a favorable attitude towards computer technology for teaching and learning. CPI provided classroom management skills to establish an environment supportive of computer use by 25 to 30 children in one classroom with only one to four computers. The teachers in the CPI program had clear examples of how to create such an environment from their visitations with their Resident Teachers. As Rother (2005) noted, "One thing that is missing from the [technology] training equation is the development of classroom management skills" (p. 34).

Figure 1. Technology oriented professional develoment lens

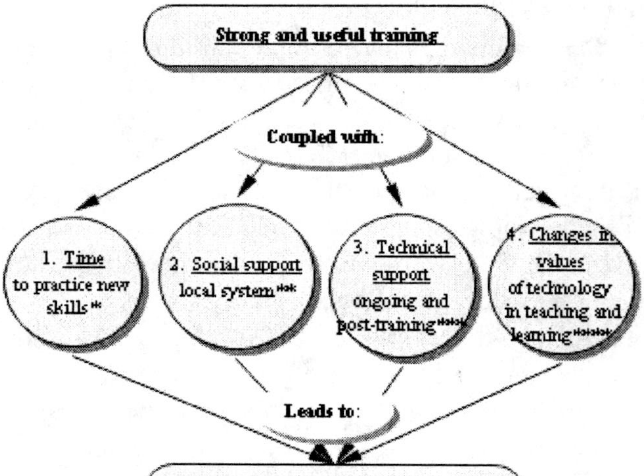

* Time means: time to practice new skills, to develop interest, to evaluate, to try skills out with students, and to adopt or reject training.
**Social support: includes Principal support of technology uses, ongoing communication through a local social support system.
***Technical support: includes ongoing post-training classroom or school support of technology.
****Change in values: includes change in classroom structures, roles and behaviors, knowledge and understanding, and uses of technology in classrooms.

From this analysis of the CPI professional development program using the six frames of reference, a composite lens or framework is proposed for predicting short- and long-term success of such efforts (see Figure 1). Strong and useful technology-oriented professional development programs, developed using the research literature as a guide and demonstrating the following principles, may be more likely to lead to sustained outcomes in classroom practice over time by providing:

1. Time to practice, to develop interest and knowledge, to evaluate usefulness for own classroom and students, to try new skills with students, and to adopt or reject the technology based on these opportunities,
2. Effective, ongoing, post-training technical support in the classroom,
3. Ongoing communication and a local social support system, including significant support from the principal and/or other influential school staff, and
4. Changes in classroom structures, roles and behaviors, knowledge and understanding, and values of technology in classrooms.

While these principles are important for any professional development effort, they are particularly important with technology to ensure teacher comfort with, and acceptance of, the technology over time, especially the importance of the supportive principal who is willing to give teachers time and encouragement to gain the knowledge and experience needed to infuse computer technology into their classrooms appropriately. Teacher comfort has been shown to have a direct relationship with intent to use technology and a teacher's prediction that he/she will use the technology (Davis, 1993).

Henry Jay Becker (1998) reported from his research using the 1989 IEA survey data (Becker, 1994) that outstanding teaching practice was found when "computer expertise was present at the same school … training opportunities were plentiful … a social support network existed … a principal-led climate for productive use of computers [was present]… and teachers had opportunities to use computers themselves during the school day" (Becker, 1998, p. 27). The findings from the study of the CPI program, and the proposed lens for assessing technology professional development, are consistent with that of Becker's. What was unique about the CPI program was including the classroom management component of infusing computers into the classroom, which often stymied most teachers' ability to use a few computers with 30 students or more. The issue was not whether research and the literature were consistent or provided evidence of best practice, but whether those who developed technology professional development programs at the local level were aware of and followed such sage advice.

Conclusion

Computers in schools are a given in the early 21st century. While the nature of computing power will no doubt change in schools as they evolve, use of computers by teachers is predicated on several factors.

1. First of all, teachers need to have an image of how computers can and will support their teaching and their students' learning. Without such a vision, computers in those classrooms remain unused or underused as reported by Cuban (2001). To ensure teachers have such a vision, professional development training opportunities are essential. However, not just any professional development program will suffice when it comes to computers.
2. The professional development program must have strong training materials, time for teachers to practice skills and uses of applications, opportunities to share with other teachers as well as opportunities to observe master teachers successfully using computers with their students. Having the opportunity to interact with students as they use computers to complete projects provides a level of understanding of the power of computers to support learning not accessible any other way.
3. Support throughout the process is essential to be sure skills are learned and honed over time. Support is measured through the environment in the school, having a principal who values computers in supporting the curriculum, opportunities in the school to share with other teachers ideas and resources, and time to continue to evolve into a master teacher with technology.
4. Support for the technology itself is essential to be sure the technology works. As noted by Schofield (1995), teachers do not want to look foolish in front of their students or peers. If the computers are unreliable, teachers will opt for other tools to support their teaching, leaving computers underused in the classroom.
5. Lastly, visible changes in values are essential in demonstrating how computers in classrooms change classroom structures (not just relegated to the back of the classroom, but into computer centers for use in strategic moments), roles and behaviors (e.g., the teacher as a guide to learning, not as the primary source of information), of knowledge and understanding of how computers can support the curriculum and different students learning needs (Gardner, 1999).

When these factors are present, in concert with a program director who sees this vision for professional development, sustained classroom change in practice by teachers and students will more likely occur. And when classroom practices change in support of educational initiatives, students gain.

Implications for Future Programming Similar to CPI

Considering the lens presented, there are further implications for a program like CPI in classrooms categorized as early childhood, elementary, or middle school. From these examples and the experience of the researcher in studying the impact of CPI on classroom teachers, infusing computers into classroom practice appears to occur more easily in both early childhood classrooms and elementary classrooms, but with more difficulty in middle grade classrooms. Middle grade teachers often identify with their subject matter first, whether teaching in an elementary or middle school. If computers have not been an integral part of teaching the English, math or science curriculum before technology is introduced through professional development, these teachers are less likely to embrace the addition of computers into their teaching practices. Additionally, if the teacher does not view computers as a link to the future, they are less likely to embrace change in teaching practices. School districts and schools must carefully consider how computers are introduced to teachers to ensure the greatest success in training programs.

As noted through the change theories used as analysis tools in this book, if teachers do not see a new strategy or technology as providing better options or opportunities for them and/or their students, they are not likely to welcome the change in practice. There were few examples in this book where teachers did not fully embrace computers into their classrooms. These teachers did not enter the CPI training opportunity because they saw an opportunity to bring new options to bear on their classroom practices, but to appease another teacher in the school. Assuring that teachers who participate in professional development programs like CPI are genuinely interested in changing classroom practices may not be possible or practical; however, if a school or school district administrators invest precious funds in training individuals, they want to know that the investment has an impact on classrooms.

In addition, there were a few examples where teachers who wholeheartedly adopted computers into their classrooms, but later receded to minimal or no use of those computers with their students. These fell into two categories—where the infrastructure did not support the use of multiple computers in the classroom, making it difficult for more than one student at a time to have access to technology, or where the computers were removed from classrooms for a school wide initiative. The infrastructure supporting uses of electrical devices in classrooms is not an easy problem to fix, but needs to be considered when a school district tries to meet demands in schools. The E-Rate program facilitated bringing high-speed Internet connections into schools and classrooms, but was not a source of funding to solve the other infrastructure problems present in older school buildings. Where there are competing demands on resources, the impact of removing resources from classrooms should be considered and addressed, realizing that all demands for resources and supplies cannot be met in all schools. These are difficult policy decisions that should not be considered in a vacuum.

Future Research

One area of research not included in this study was how administrators perceive the use of computers in supporting teaching and learning in classrooms, or their perception of CPI. Many teachers voiced the support of their principal in their participation in the CPI program. Principals were not interviewed for this book since teacher voices were considered most important. However, future research on professional development opportunities around infusing computers into classroom practices should include the principals' voices.

Professional development opportunities for teachers abound. With the NCLB legislation, teachers must meet requirements to be highly qualified teachers, meaning they will be seeking additional training (U.S. Department of Education, 2001). Where schools and school districts offer opportunities for teachers to learn more about infusing computer solutions into classrooms, careful attention should be paid to how these programs actually impact classroom practices. Additional longitudinal research is needed to demonstrate the long term impact of these programs on teacher practices, and ultimately on student learning. The true test of professional development is how students demonstrate knowledge through classroom assignments as well as both low- and high-stakes testing required by NCLB.

References

Ball, D. L. (1996). Teacher learning and the mathematics reforms: What we think we know and what we need to learn. *Phi Delta Kappan, 77*(7), 500-508.

Becker, H. J. (1994). How exemplary computer-using teachers differ from other teachers: Implications for realizing the potential of computers in schools. *Journal of Research on Computing in Education, 26*(3), 291-321.

Becker, H. J. (1998). Running to catch the moving train: Schools and information technologies. *Theory into Practice*, *37*(1), 20-30.

Berman, P. and McLaughlin, M. W. (1978). *Federal programs supporting educational change, Volume VIII: Implementing and sustaining innovations*. Santa Monica, CA: Rand.

Cuban, L. (2001). *Oversold and underused.* Boston: Harvard University Press.

Davis, F. D. (1993). User acceptance of information technology: System characteristics, user perceptions and behavioral impacts. *International Journal of Man-Machine Studies, 38*, 475-487.

Docterman, D.A. (1998). *Great teaching in the one-computer classroom.* Fifth edition. Watertown, MA: Tom Snyder Productions.

Dwyer, D. C., Ringstaff, C., & Sandholtz, J. H. (1990). The evolution of teacher's instructional beliefs and practices in high-access-to-technology classrooms. Paper presented at the annual meeting of the American Educational Research Association, April, Boston.

Fisher, Dwyer, D. C., & Yocum, (1996). *Education and technology: Reflections on computing in classrooms.* San Francisco: Josey-Bass.

Fullan, M. (2001). *The new meaning of educational change.* Third Edition. New York: Teachers College Press.

Fullan, M. & Pomfret, A. (1977). Research on curriculum and instruction implementation. *Review of Educational Research, 47*(1), 355-397.

Gardner, H. (1999). *Intelligence reframed: Multiple intelligences for the 21st century.* New York: Basic Books.

Gordon, A. R. (1996). *Computers in the classroom: How teachers and students are using technology to transform learning.* San Francisco: Jossey-Bass Publishers and Apple Press.

Guskey, T. R. (2000). *Evaluating professional development.* Thousand Oaks, CA: Corwin Press.

Hord, S. M., Rutherford, W. L., Huling-Austin, L., & Hall, G. E. (1987). *Taking charge of change.* Alexandria, VA: ASCD.

Joyce, B. and Showers, B. (2002). *Student achievement through staff development.* Third edition. Alexandria, VA: ASCD.

Little, J. W. (1989). District policy choices and teachers professional development opportunities. *Educational Evaluation and Policy Analysis, 11*(2), 165-179.

President's Committee of Advisors on Science and Technology, Panel on Educational Technology. (1997). *Report to the President on the Use of Technology to Strengthen K-12 Education in the United States.* As cited in A. Cattagni, & E. Ferris (2001), Internet access in U.S. public schools and classrooms: 1994-2000. Washington, D.C.: U.S. Department of Education, National Center for Educational Statistics, Fast Response Survey System, p. 3.

Rogers, E.M. (1962, 1972, 1983, 1995, 2003). *Diffusion of innovation.* Fourth edition. New York: The Free Press.

Rother, C. (2005). Is technology changing how you teach? *T.H.E. Journal, 33*(3), 34-36.

Sharan, S. and Hertz-Lazarowitz, R. (1980). A group investigation method of cooperative learning in the classroom. In S. Sharon, P. Hare, C. Webb, & R. Hertz-Lawarowitz (Eds), *Cooperation in education,* (pp. 14-46). Provo, UT: Brigham Young University Press.

Schofield, J. W. (1995). *Computers and classroom culture*. New York: Cambridge University Press.

Shulman, L. (1993). Teaching as community property: Putting an end to pedagogical solitude. *Change, 25*(6), 6-7.

U.S. Department of Education. (2001). *No Child Left Behind Act.* Washington, D.C.

Wilson, S. M. and Berne, J. (1999). Teacher learning and the acquisition of professional knowledge: An examination of research on contemporary professional development. In A. Iran-Nejad, & P.D. Pearson (Eds), *Review of Research in Education*, Volume 24, pp. 173-209.

Endnote

1 For more information about the E-Rate program, see the report by Cattagni and Ferris (2001) last accessed on August 3, 2006 from http://nces.ed.gov/surveys/frss/publications/2001071/

Appendix

Trademarks, Registered trademarks, and Service Marks Found in this Book

America Online, CompuServe, and AOL are registered trademarks of AOL LLC, in the United States and/or other countries.

ACOT, Apple, Apple IIGS, Appleworks, iMovies, iPod, iMac, iBook, eMac, Mac, MacOS, Macintosh, and Hypercard are registered trademarks or service marks of Apple, Inc. in the United States and/or other countries.

ClarisWorks for Kids is a registered trademark of Claris Corporation, a subsidiary of Apple Computers.

Ask Jeeves for Kids is a trademark of IAC Search and Media in the United States and other countries.

Accelerated Reader is a trademark of Renaissance Learning™, Inc.

BrainPOP is a registered trademark of BrainPOP, LLC.

Carmen SanDiego, KidPix, Learning Village, PrintShop, Tesselmania, and Wordstar and registered trademarks of Riverdeep®, Inc. of the United States and/or other countries.

Compaq is a registered trademark of Hewlett-Packard Development Company, LP, Texas.

DIBELS is a registered trademark of Dynamic Measurement Group, Inc. Enchanted Learning and Little Explorers are trademarks or registered trademarks of Enchanted Learning, LLC.

Eudora is a registered trademark of QUALCOMM Incorporated.

First in Math is a registered trademark of Suntex International Inc.

Google is a trademark of Google, Inc.

Harvard Graphic is a registered trademark of Serif Incorporated in the United States and/or other countries.

HyperStudio is a registered trademark of The Software MacKiev Company.

Kodak is a registered trademark of the Eastman-Kodak Company.

IBM, PC, XT, AT, Lotus, Lotus Notes are registered trademarks of IBM Corporation in the United States and/or other countries.

Intel, 80386, and 80486 are trademarks or registered trademarks of Intel Corporation of the United States and/or other countries.

Inspiration and Kidspiration are registered trademarks of Inspiration Software®, Inc.

ISTE is a registered trademark of the International Society for Technology in Education in the United States and/or other countries.

ISpy is the trademark of Jean Marzollo and Walter Wick.

Lego is a registered trademark of the Lego Group in the United States and/or other countries.

LoTi is the registered trademark of Learning Quest, Inc.

Microsoft, Excel, MS-DOS, and PowerPoint are registered trademarks of Microsoft Corporation in the United States and/or other countries.

PLATO is a registered trademark of Plato Learning, Inc, in the United States and Canada.

Rand is a registered trademark of the Rand Corporation in the United States and/or other countries.

Reading Blaster is a registered trademark of Knowledge Adventure®.

Read180 and Tom Snyder Productions, Inc. are registered trademarks of Scholastic, Inc.

School House Rock is a registered trademark of the Disney Corporation.

Tandy is a registered trademark of RadioShack™ Corporation.

T.H.E. Journal is a trademark of 1105 Media, Inc.

WordPerfect is a registered trademark of Corel® Corporation in Canada, the United States, and/or other countries.

Yahooligans is a registered trademark of Yahoo®, Inc.

WYNN is a trademark of Freedom Scientific®, Learning Systems Group.

Zip is a registered trademark of Iomage® Corporation in the United States and/or other countries.

100 Book Challenge is a registered trademark of the American Reading Company.

About the Author

Catherine Schifter, PhD is an associate professor in Curriculum, Instruction and Technology in Education at Temple University, USA, a Carnegie Scholar (2000-2001). She was the director of the Temple Online Learning Program from 1996 through 2000. From 2002 through 2004, Dr. Schifter established the Temple University Teaching and Learning Center to support faculty in all endeavors of teaching and learning in traditional and online environments. She is the chair of her department beginning in fall 2007. Recent scholarship includes reviewing national trends for distance education faculty support models in higher education, and evaluation of staff development programs for infusing technology into classrooms in K-12 education. Her teaching responsibilities include infusing technology into teaching into K-12 classrooms, overseeing of the senior seminar, teaching overall research design as well as qualitative research methods.

Index

M

N

O

P

R

S

T